RENOVATED SPACES

New Life for Old Homes

RENOVATED SPACES

New Life for Old Homes

Àlex Sánchez Vidiella
Francesc Zamora Mola

LOFT

Renovated Spaces. New Life for Old Homes

Editorial coordinator: Simone K. Schleifer
Assistant editorial coordinator: Aitana Lleonart
Editors & texts: Àlex Sánchez Vidiella, Francesc Zamora Mola
Art director: Mireia Casanovas Soley
Design and layout coordination: Claudia Martínez Alonso
Layout: Cristina Simó, Concepción Papió

Loft Publications, S.L.
Via Laietana 32, 4º, of. 92
08003 Barcelona, Spain
P +34 932 688 088
F +34 932 687 073
loft@loftpublications.com
www.loftpublications.com

ISBN: 978-84-92463-57-2

Printed in Spain

>INDEX

>INTRODUCTION

Many are the reasons for renovating, for instance to add comfort or increase utility, not to mention investment that adds value to the property. Home renovations may incorporate extensions as the simplest and most efficient way to add living space to a home. For example, the creation of additional space may satisfy the needs of an expanding family or provide accommodation for visitors. Other reasons might be the addition of storage space or rooms for special purposes such as bathrooms and office space. Also, the reconfiguration of the house layout may improve circulation and orientation to sun and views.

Renovations are often done to reanimate an existing construction that, due to external factors, has become unpractical if not obsolete. They offer opportunities for adaptation on different levels: change in the use of space, rearrangement of the spatial layout that may affect the relation between the different rooms in the building or the relation between interior and exterior, or an update of the atmosphere of the existing spaces.

We renovate and upgrade living spaces with an effort which might range anywhere from just painting and wall covering, to removing and reconstructing walls and enlarging the building. As a consequence, some spaces may have been subject to multiple layers of modifications through their history. In buildings encrusted with these accumulated modifications over time are recorded the superimposed cycles of changes in use, and technologies that have progressively transformed the original structure.

This book intends to analyze a selection of projects based on the level of transformation.

At one end of the spectrum, transformation may consist of stripping down to the bare bones and starting anew. What remains is simply used as a container holding in new elements. This is the case of a former smithing plant in the outskirts of Brussels featured here. The existing structure was subject to substantial demolition, and left as shell for the architectural interventions

that resulted in the creation of live and workspaces. This conversion is an important transformation that not only affects the layout and the appearance of the building but also its use. In cities everywhere, changes in land use are opening up formally industrial or commercial zones to residential use, making these structures eagerly sought after as places to explore and enjoy as living spaces, and offering an interesting alternative to otherwise more conventional living.

At a different level, transformation allows some of the existing fabric of the building to become an important element in the remodel, creating a dialog between old and new. An example of this is "The Old Granary" project included in this book that consists of two stone barns converted into high quality accommodation. The scope of work included extensive alterations to create an open plan living space while emphasizing the contrast between original construction and new elements.

Finally, at the other end of the spectrum, transformation simply affects a superficial layer. In this case, the original layout of the space is maintained but minor alterations may occur. The project in Barcelona's 22@ neighborhood was done on a very tight budget. Technicolor inspired fields and stripes of saturated tints are the simple but effective theme of this design. The plan is little changed: only the wall that enclosed the kitchen is demolished to create a spacious living area.

This book explores through different projects the aesthetic directions and technical solutions that architects and designers have followed to adapt to the occupant's needs.

The volume includes a worldwide selection of works in which both variants — the creative and the pragmatic — have been comprehensibly explored by means of sketches, diagrams, technical drawings, working models and photographs taken before, during and after the remodel. All of this material helps illustrate the renovation process from concept to completion resulting in examples of the cohabitation of two complementary parts: the old and the new.

FARM HOUSE VOGES REDUX

> Wennigsen, Germany | 2008 | Duration of project: 15 months | 4,477 sq ft | © Olaf Baumann <

Despang Architekten

This project came about as the result of research conducted by the architects' studio into ways to improve the most traditional type of north German housing, which characteristically has a timber structure. This material is used sparingly, and the gaps between the timbers are filled with a conglomerate of various materials including earth, mud and reeds.

The main requirements in the renovation brief of this two-story building were, firstly, to create a more open façade to enhance daylighting and, secondly, to supply an alternative means of energy. The living space on the ground floor, in the center of the building, is small and consists of a thermal zone that is separated from the outer structure by means of a triple-layered glazed wall that allows daylighting to enter.

Opening the ground floor to the exterior did not alter the typical features of a north German home. The timber superstructure was conserved, and a porch was created between this and the new thermal glazed space. This porch serves to protect inhabitants from the elements.

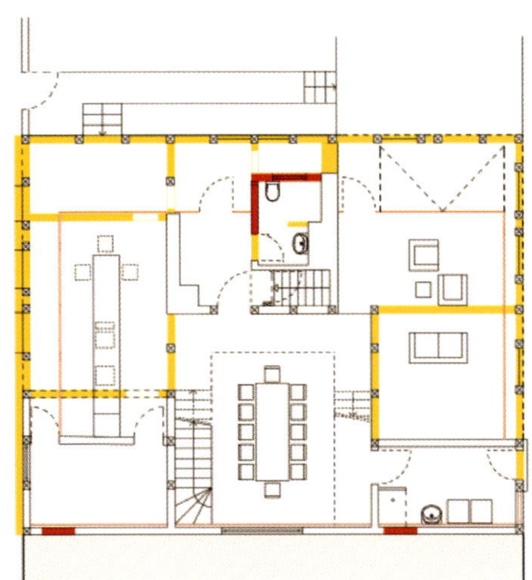

Existing ground floor plan

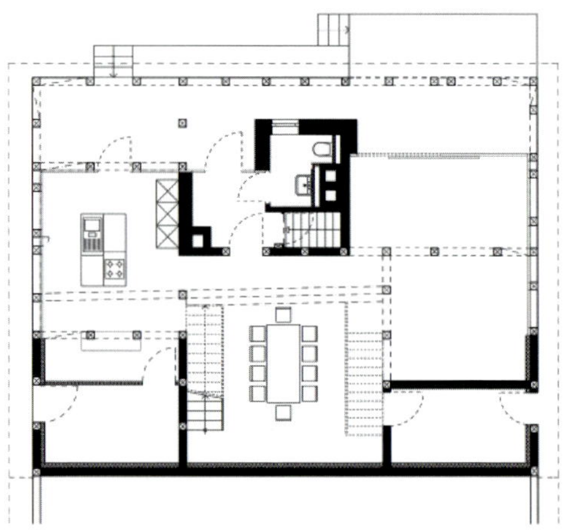

New ground floor plan

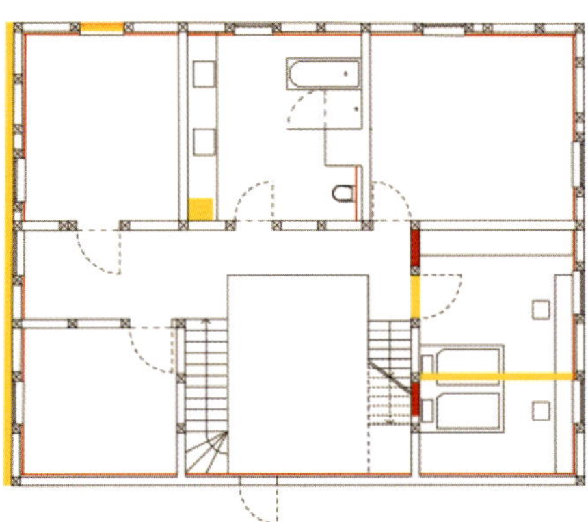

Existing second floor plan

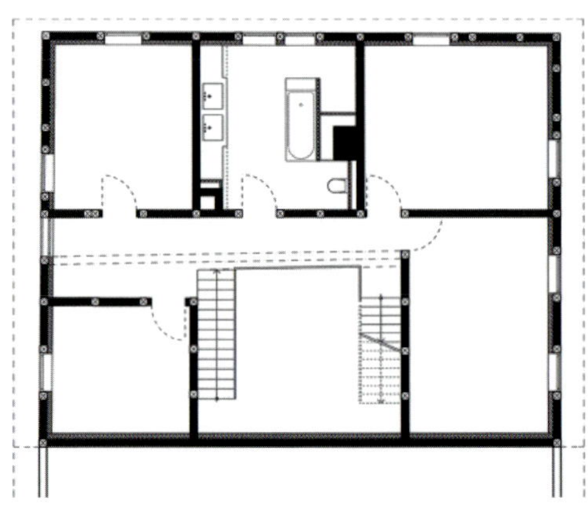

New second floor plan

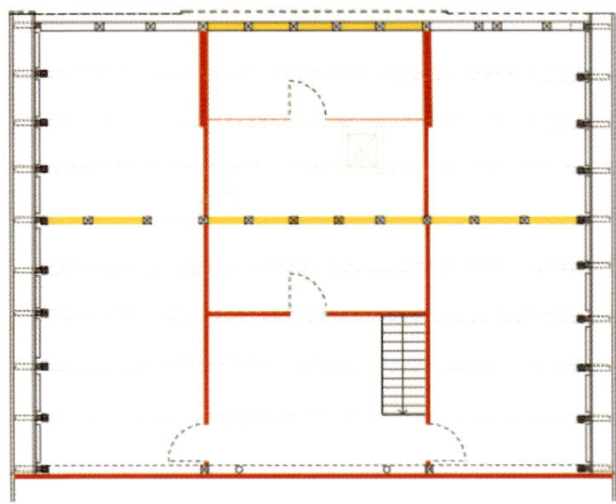

Existing third floor plan

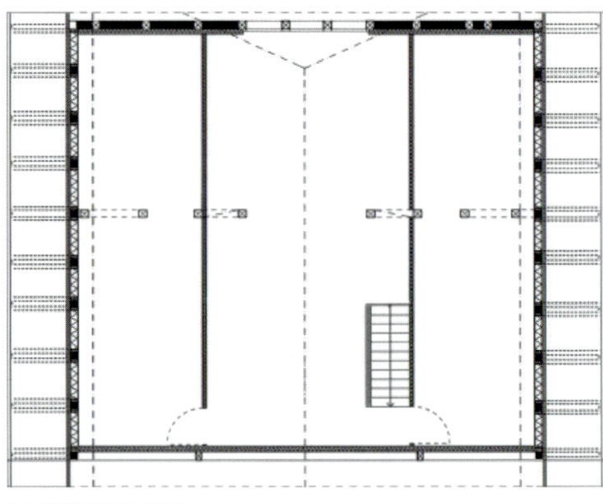

New third floor plan

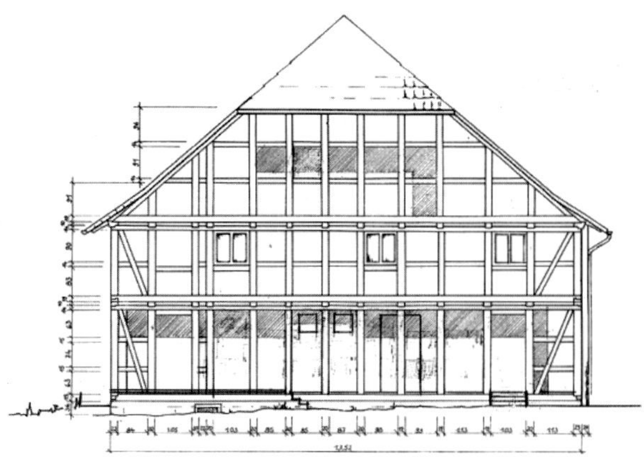

North elevation

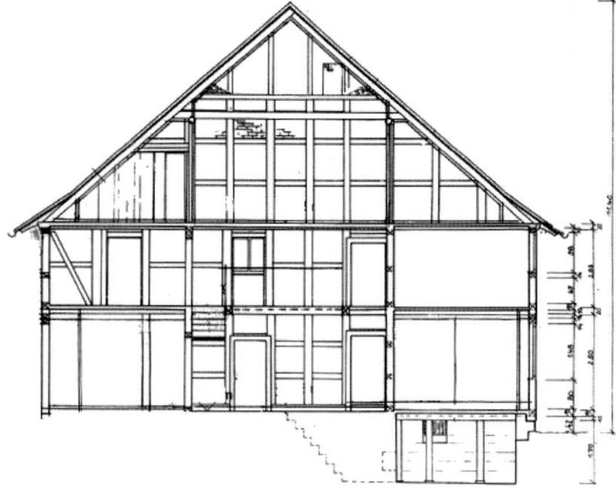

Section 1

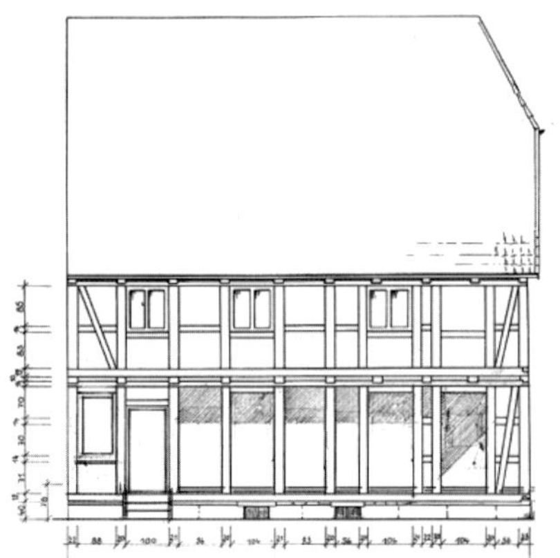

East elevation

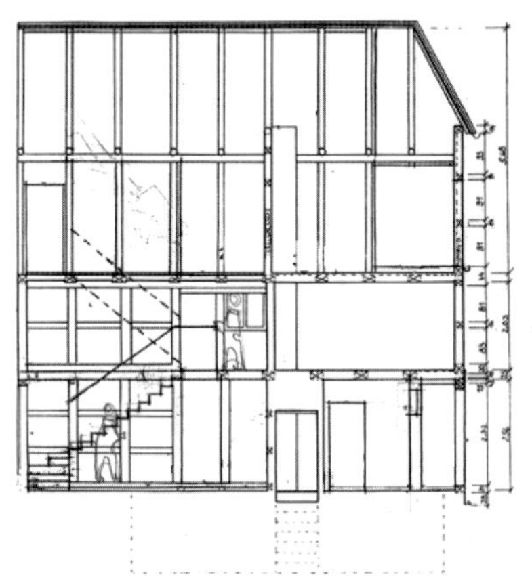

Section 2

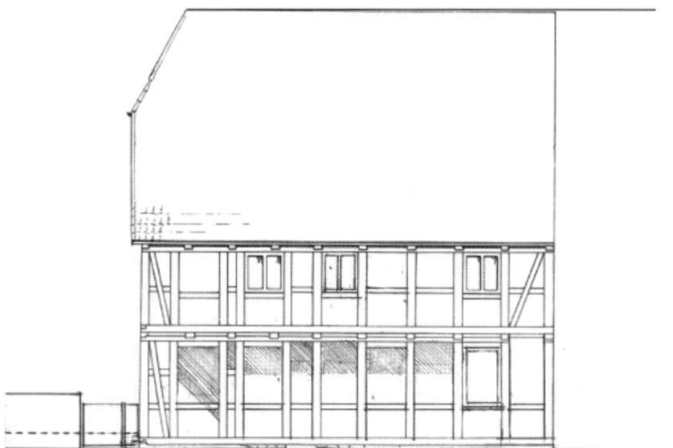

West elevation

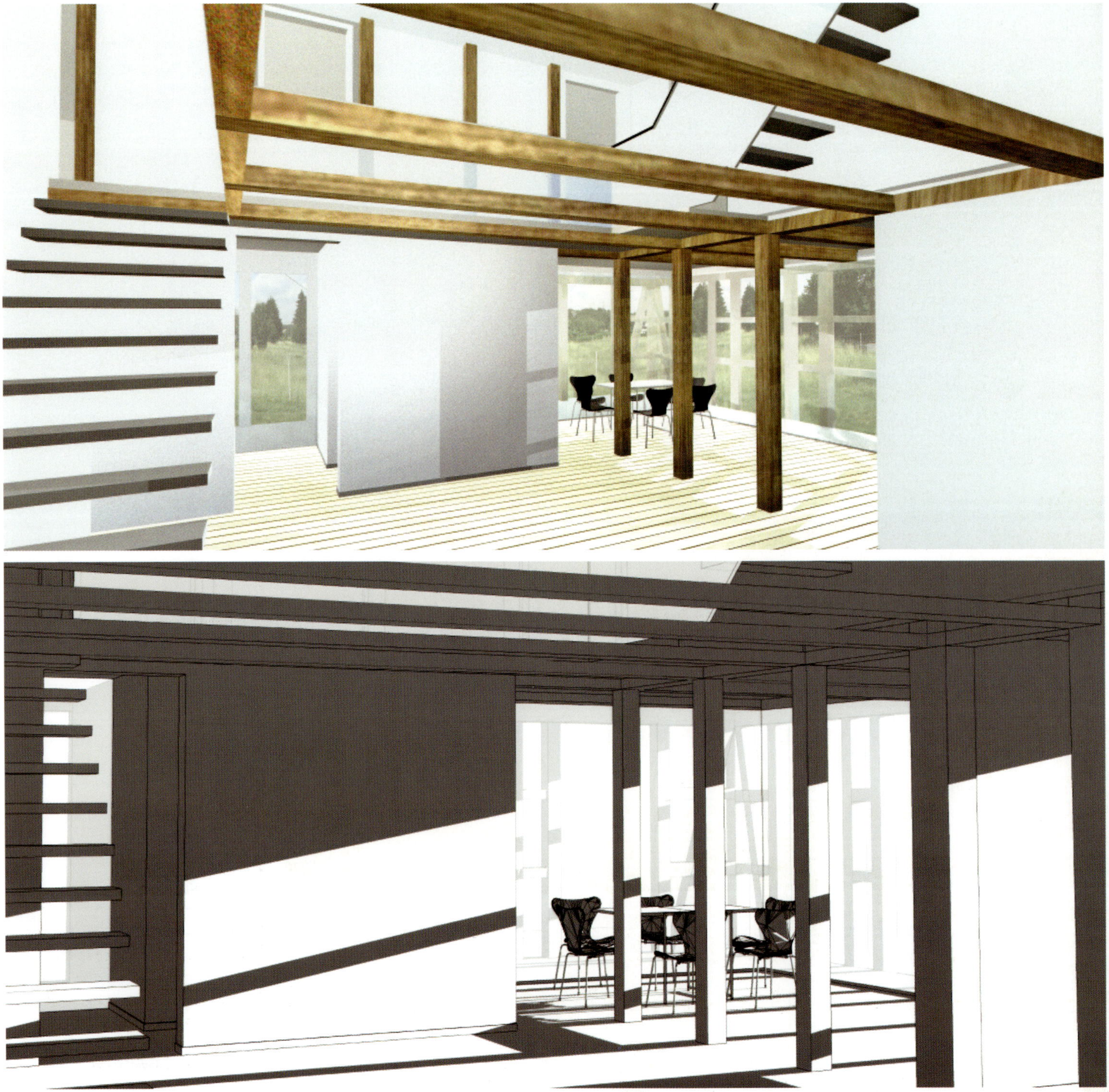

Light studies

The architects decided against opening the second level up to the exterior, since the more private living spaces are located on this level. This decision therefore created an architectural hierarchy whereby the common spaces are located on the lower level and the private spaces on the level above.

Piero Camoletto, Luca Rolla

TOWER HOUSE

> Bee, Italy | 2006 | ·Duration of project: 5 months | 614 sq ft | © Andrea Martiradonna <

This mid-19th century tower is located in the old town in Bee, a town close to Lake Maggiore. The building was in urgent need of work, as it was very run down.

The roof terrace was demolished and rebuilt. The wood frames and glass in the wall openings were removed to reinforce and protect the local stone walls from damp. The first floor, which is partially sunken due to the lot's slope, was raised and ventilated. The joists of the wood flooring were replaced by steel as the floor was raised. The new flooring solutions used consisted of wood and semi-transparent elements.

Existing third floor plan

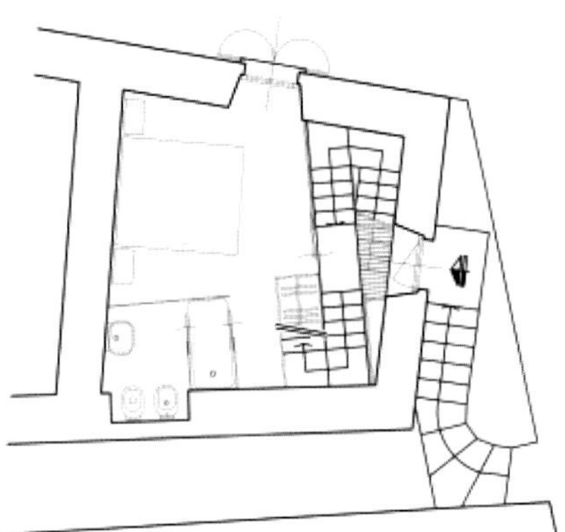

Existing second floor plan

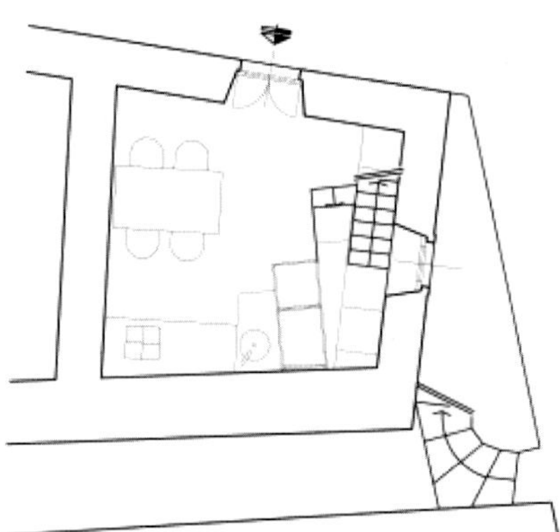

Existing ground floor plan

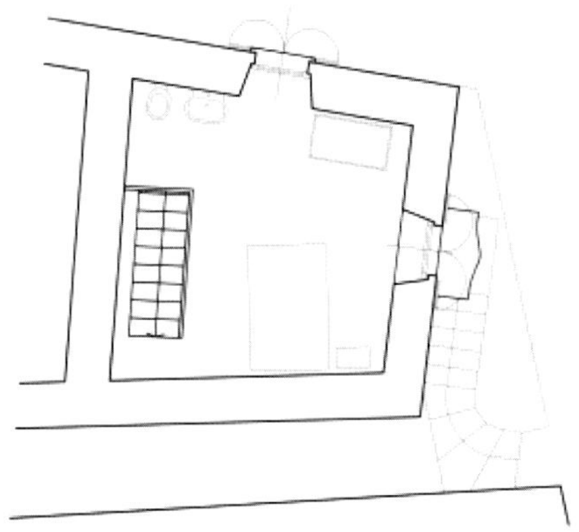

New third floor plan

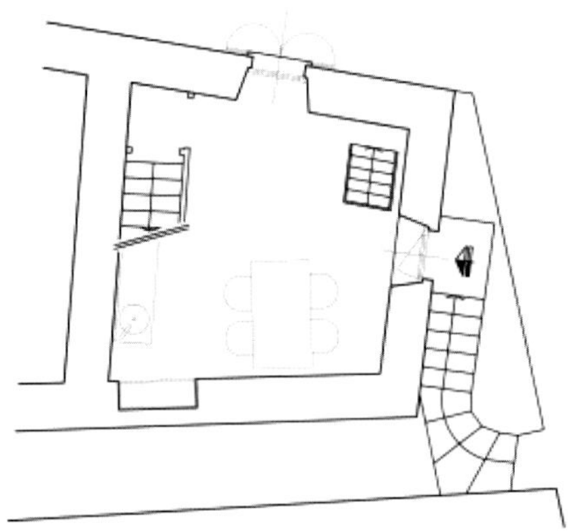

New second floor plan

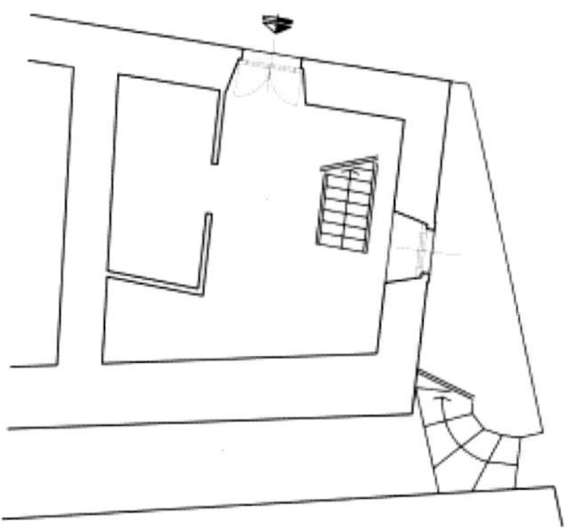

New ground floor plan

Pre-existing facades

New facades

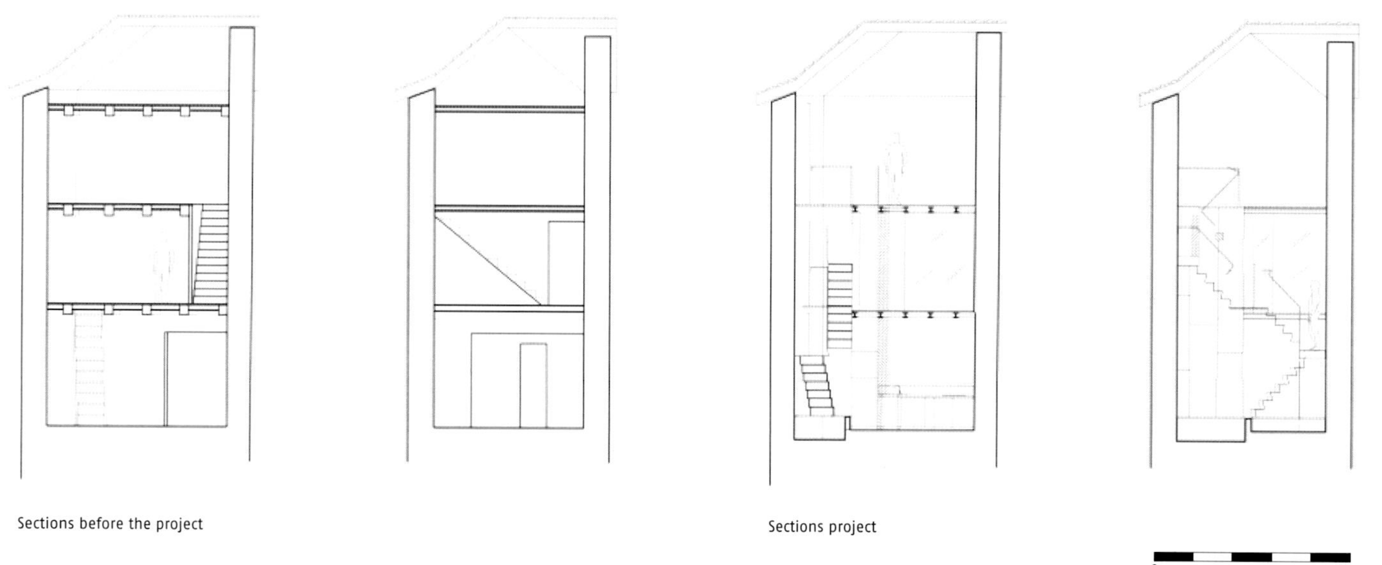

Sections before the project

Sections project

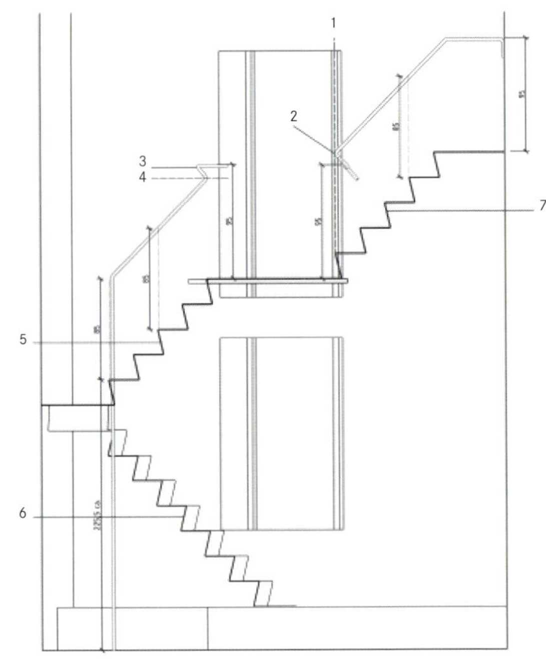

1. Line of tread
2. Aligned with line of first tread
3. Intersection between center line of handrail and side of ramp 1
4. Topline of elevation +85 above landing
5. Ramp 2
6. Ramp 1
7. Ramp 3
8. Center line of handrail paralel to tread edge
9. Center line of handrail paralel to tread edge
10. Ramp 1
11. Tigth to ramp
12. Ramp 2
13. Paralel to ramp
14. Aligned with ramp
15. Paralel to ramp
16. Ramp 3
17. Ramp 4

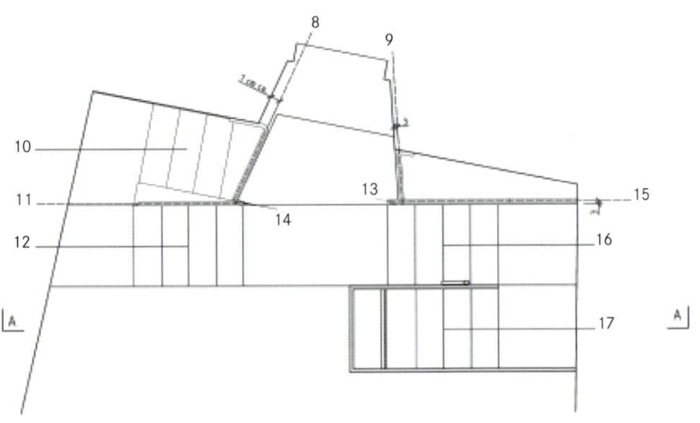

Stair plan and section

The building has three spaces located on three different levels. The levels are connected by a fine Cor-Ten steel staircase. When installing the staircase, this material enabled it to be molded and rotated to fit in the space. The staircase is suspended by braces inset in the frame.

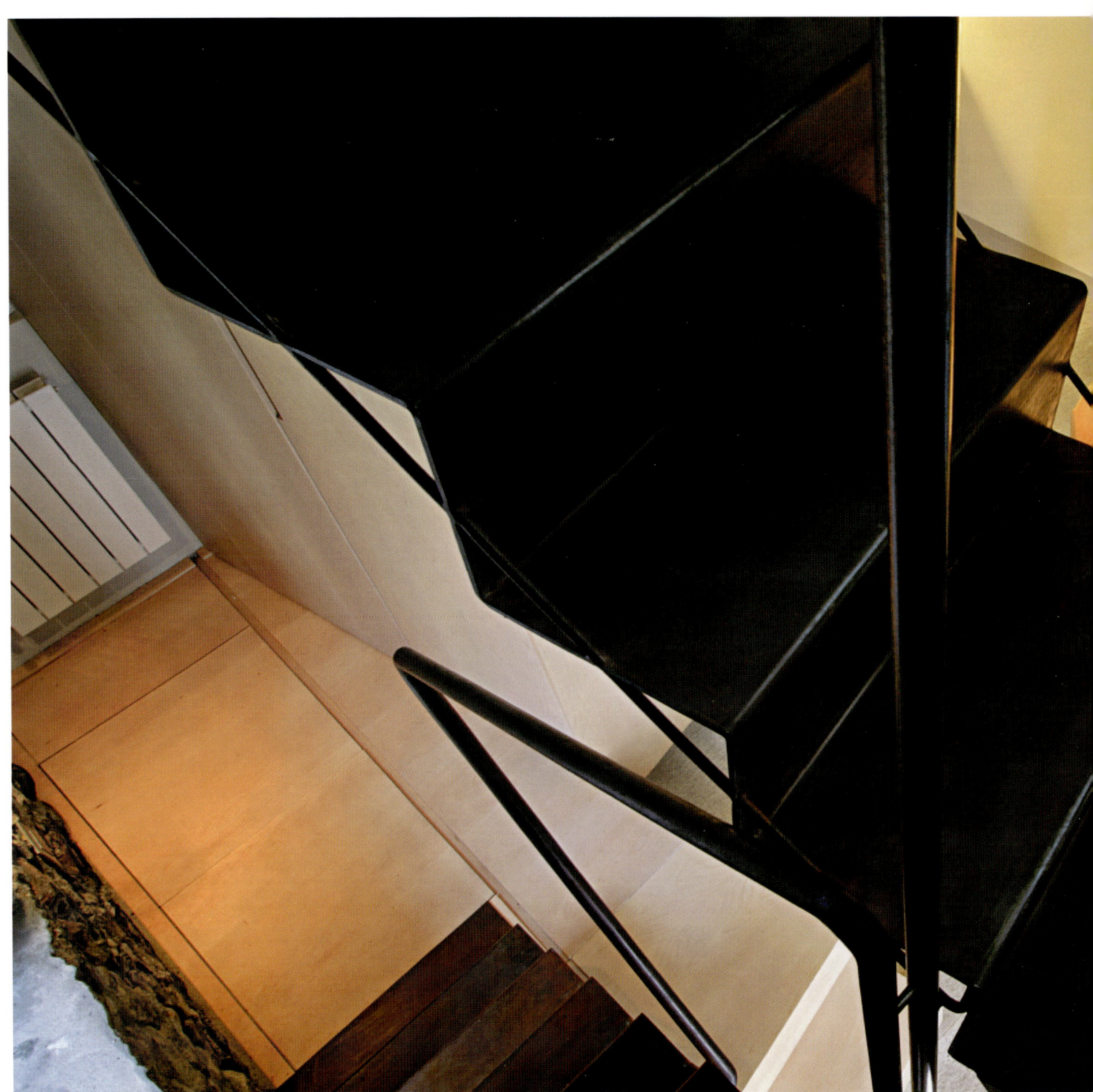

The architects opted for different flooring solutions: stone was used on the ground floor since it is harder wearing, while bleached larch was used on the other levels. The relation between the old and the new is patent in the dwelling's walls.

Mattinson Associates

> Isle of Wight, United Kingdom | 2008 | Duration of project: 20 months | 26,127 sq ft | © Carlos Domínguez <

This conversion program encompassed a house, an annex, and an open-air garage, all owned by Pedvin, a real estate developer. The property also had two stone granaries that were converted to provide greater comfort. As well as the radiant floor heading, double-glazed windows framed with fine hardwood were installed in the windows and doors. The views from the buildings overlook the rolling terrain, with a garden and lake.

The design concept was to create a vibrant space, connected to the exterior and which emphasized the interaction between the old and new features. The idea was to maintain the main natural materials and the earthy, neutral colors. This was the case for the roof, where the original beams of the granaries were reused. Likewise, big windows were added to provide spectacular views of the surrounding landscape, emphasizing the relationship between the interior and exterior spaces.

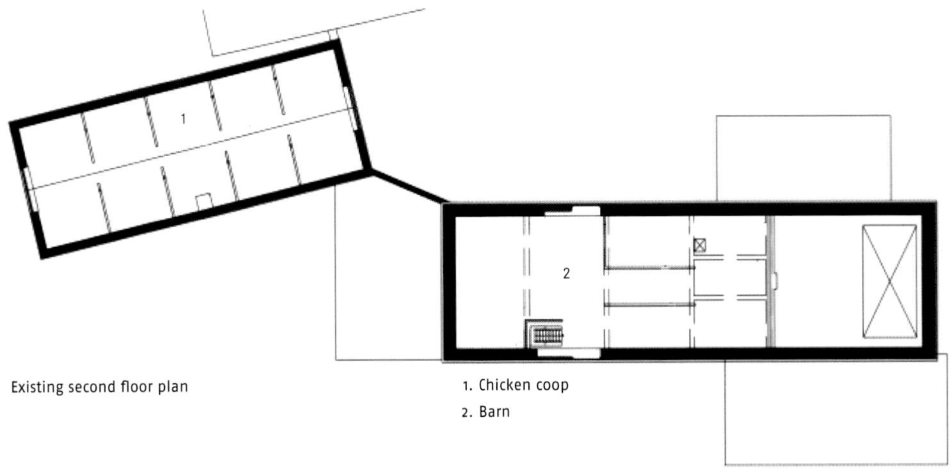

Existing second floor plan

1. Chicken coop
2. Barn

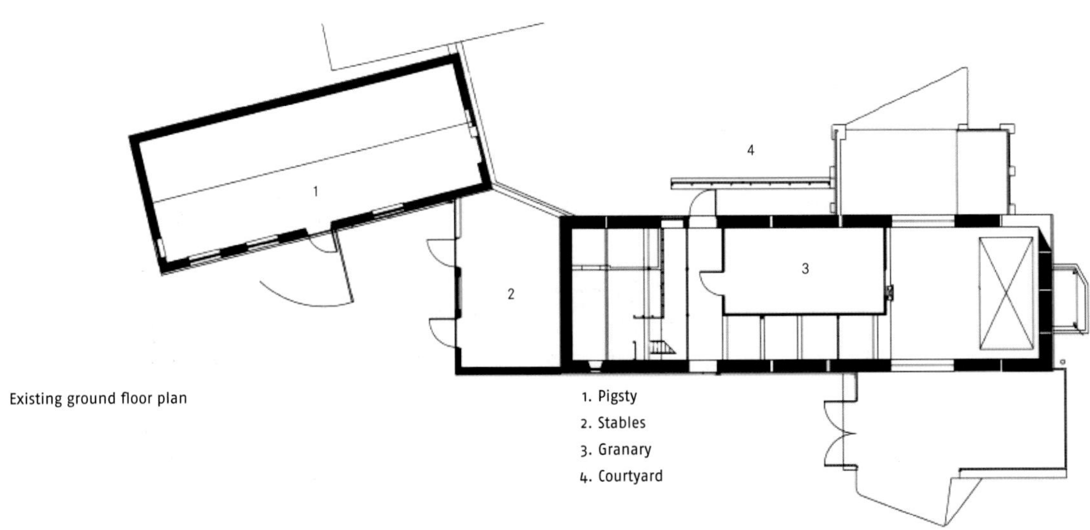

Existing ground floor plan

1. Pigsty
2. Stables
3. Granary
4. Courtyard

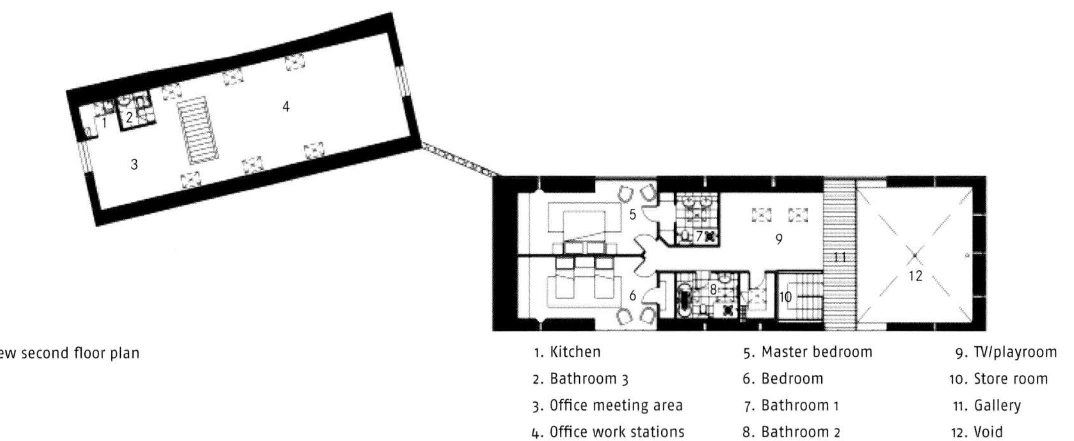

New second floor plan

1. Kitchen
2. Bathroom 3
3. Office meeting area
4. Office work stations

5. Master bedroom
6. Bedroom
7. Bathroom 1
8. Bathroom 2

9. TV/playroom
10. Store room
11. Gallery
12. Void

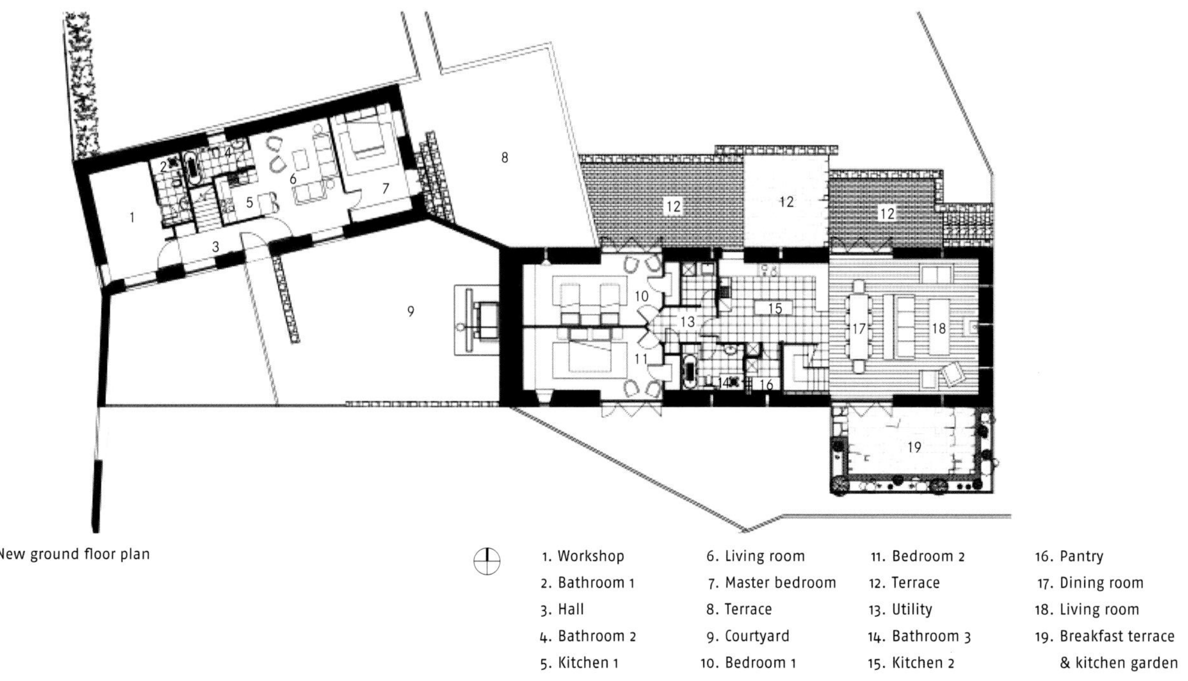

New ground floor plan

1. Workshop
2. Bathroom 1
3. Hall
4. Bathroom 2
5. Kitchen 1

6. Living room
7. Master bedroom
8. Terrace
9. Courtyard
10. Bedroom 1

11. Bedroom 2
12. Terrace
13. Utility
14. Bathroom 3
15. Kitchen 2

16. Pantry
17. Dining room
18. Living room
19. Breakfast terrace & kitchen garden

The original roof was completely removed, although the timber planks and beams were conserved and restored. The roof was finished in a natural Welsh slate, and Ventnor Green stone was recovered to be used in the new constructions.

The remodeled complex consists of a main house with four bedrooms, a lounge, a TV and games room, three bathrooms, a kitchen, a pantry, and a storeroom; an annex with a hall that leads to a workshop, an office, a double bedroom, a lounge, a kitchen-dining room and a bathroom; and finally an open-air garage that can park two large automobiles, and a cycle storage area.

The reception area of the main building, which conserves and exposes the original stone walls, is completely open, creating a spectacular impression from the living room. The two levels of the house have been divided using a balustrade, which does not interrupt this global vision of light. The new oak floor does not visually distort the travertine paving. With respect to the lighting, there is daylighting obtained through the large windows, and artificial lighting, provided by art deco-style lamps.

The bedrooms located on both levels have extensive views of the surrounding landscape and are large enough to contain different areas. The bedrooms on the first floor open up through sliding and folding doors, and enable configurations as a single, open space in the summer months, or a closed, private space during the winter. The bathrooms have been customized and paved with travertine or natural welsh slate.

Josep Maria Esquius Prat

> El Bages, Spain | 2008 | Duration of project: 15 months | 1,830 sq ft | © Lourdes Jansana <

This 1912 property enjoys a rural location close to Mount Montserrat, Barcelona. The original building's features included: a partially sunken floor, two floors of apartments and an attic level used as storage area and containing the water tanks. The exterior finishes were very simple: cement painted white and wooden blinds. The sloping roof was inset with small skylights.

The main intervention consisted in reforming the interiors to condition the spaces and installations to meet the needs of modern living. A small volume was added on the ground floor in the same style as the rest of the house, providing more space in the kitchen. The old mosaic floors were restored and the guest washrooms were relocated.

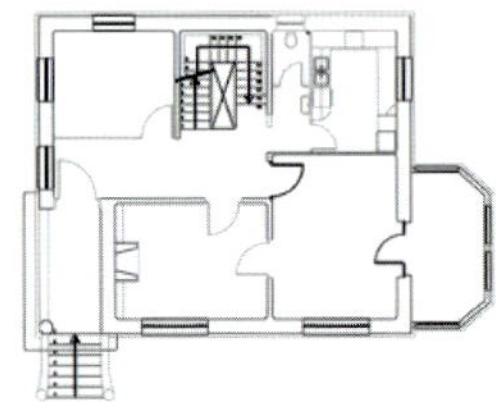

Existing ground floor plan

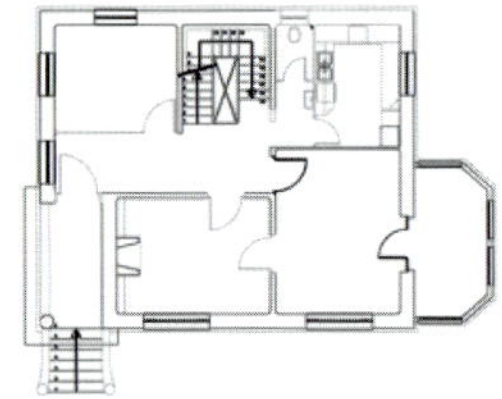

Existing second floor plan

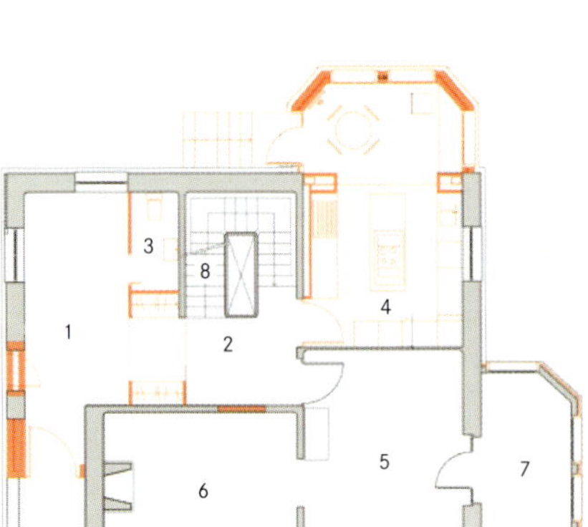

New ground floor plan

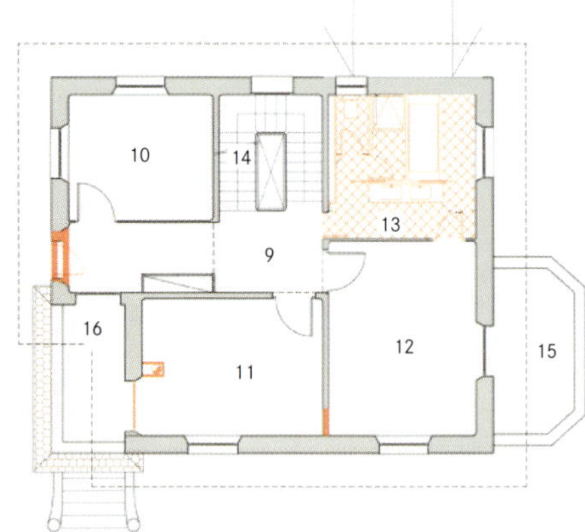

New second floor plan

West elevation

South elevation

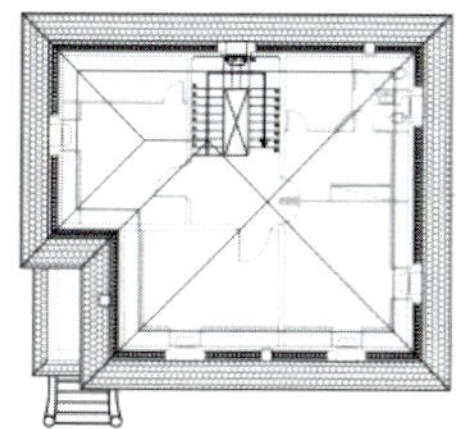

Existing third floor plan

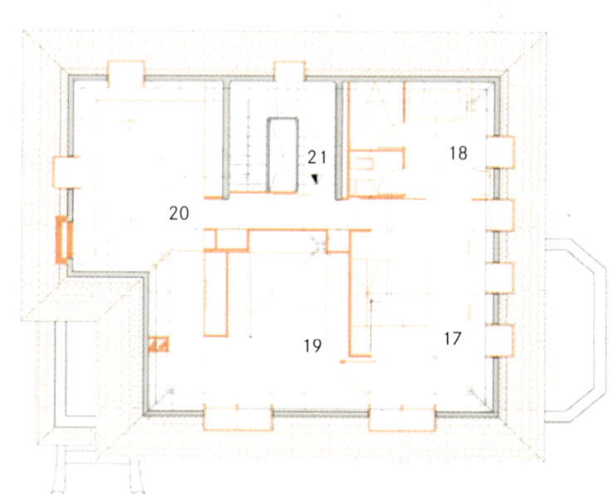
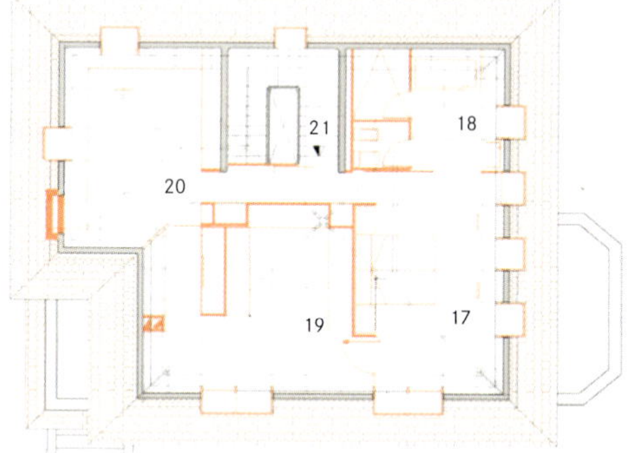

New third floor plan

1. Hall
2. Vestibule 1
3. Bathroom 1
4. Kitchen
5. Dining room
6. Living room
7. Gallery

8. Stairs 1
9. Vestibule 2
10. Bedroom 1
11. Bedroom 2
12. Bedroom 3
13. Bathroom 2
14. Stairs 2

15. Terrace 1
16. Terrace 2
17. Bedroom 4
18. Bathroom 3
19. Dressing room
20. Office
21. Stairs 3

East elevation

North elevation

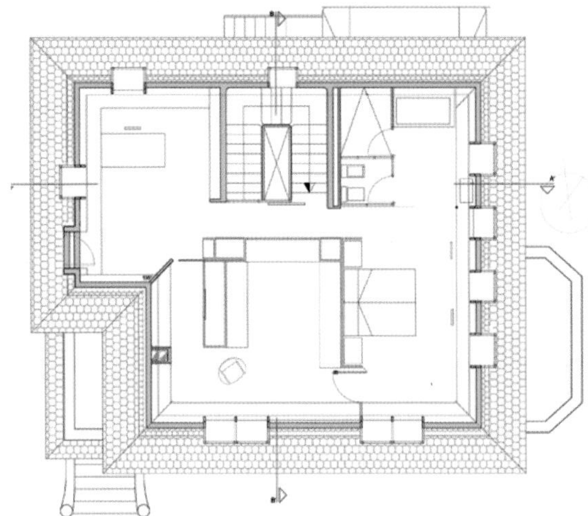

New third floor plan

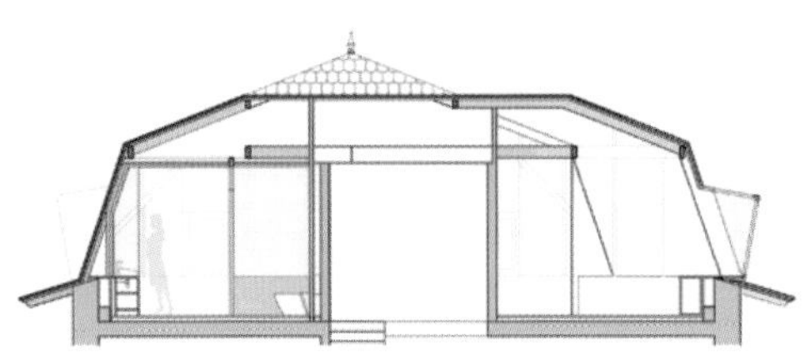

Building section

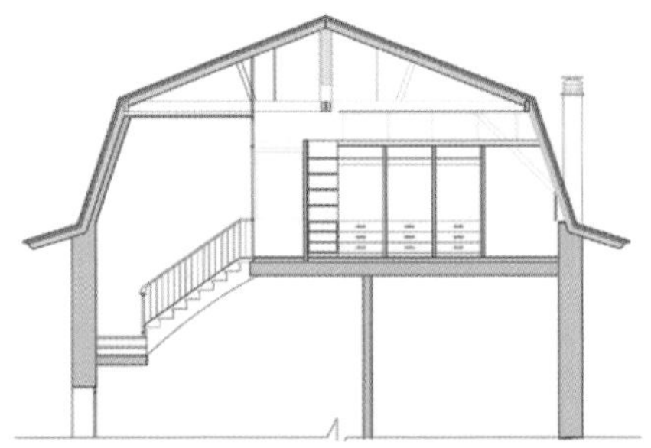

Section A–A'

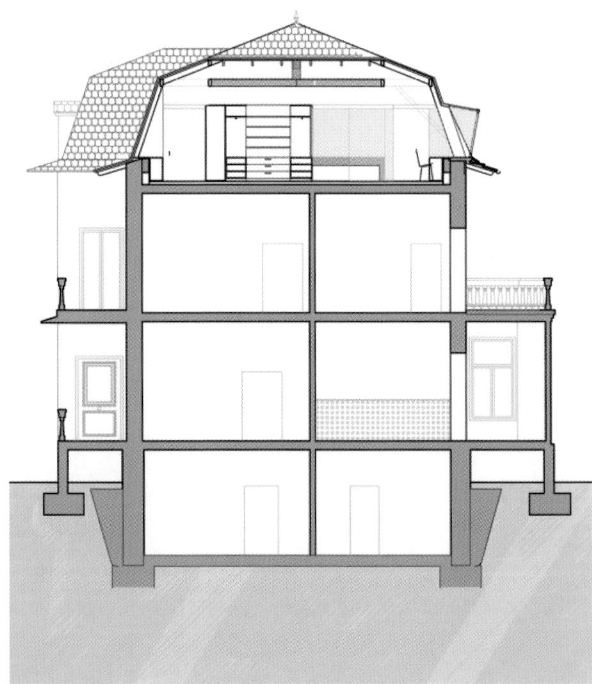

Section B–B'

The materials utilized in the renovation comprised granite, slate on the roof and in the flooring. The wooden beams with wrought iron details are an important feature in the new design. The bespoke furniture serves to divide the different spaces. The roof has flat tiles and the façade features Nídia Esquius sgraffito.

The common spaces are arranged in the first floor, while the private rooms, such as the bedrooms, are located on the second floor. The original layout of this floor was preserved, adding a kid's bathroom.

Most of the work was carried out on the uppermost level. From its former life as storage space, this level has been transformed into a unique space that consists of a bedroom, dressing room, bathroom and office, all without concealing the original roof structure, which has been left exposed.

DM35 HOUSE

> Mérida, Yucatán, Mexico | 2008 | Duration of project: 10 months | 5,619 sq ft | © Roberto Cárdenas Cabello <

Estilo Arquitectura

Over time, the DM35 home, built in the post-revolution period, had been modified on several occasions, changing its use from residential to commercial. The architectural intervention focused on recovering the original state of the façade: original ornaments were restored and extensions were demolished to recover an area to be primarily used as garden space.

Inside, the needs of the new occupants required more drastic changes. The existing spaces were adapted to take on new functions and a pool, garage, kitchen and several bathrooms and terraces were constructed. The pool, located perpendicular to the prism housing the common area, became the central feature of the outdoor space.

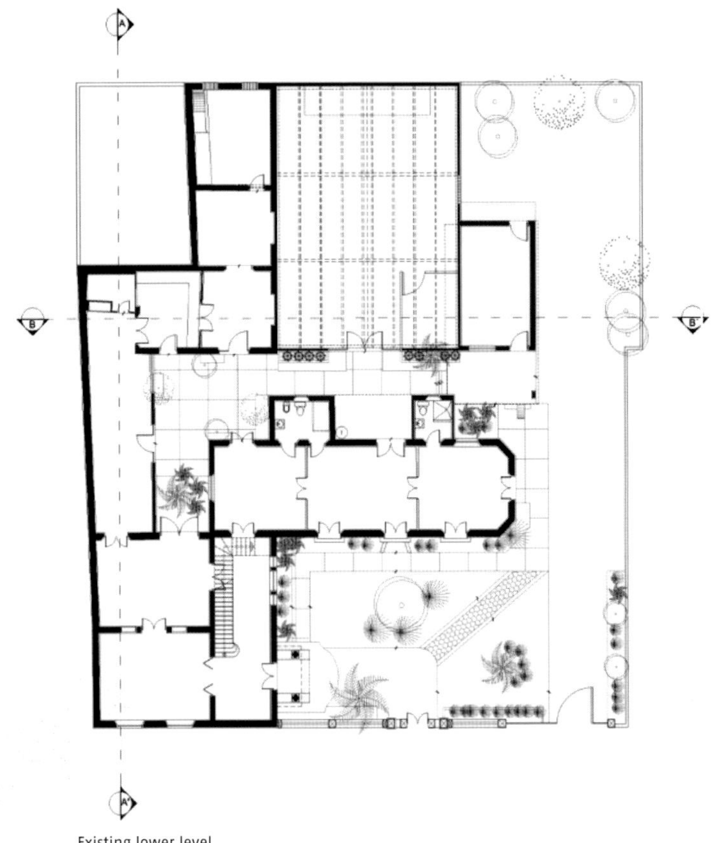

Existing lower level

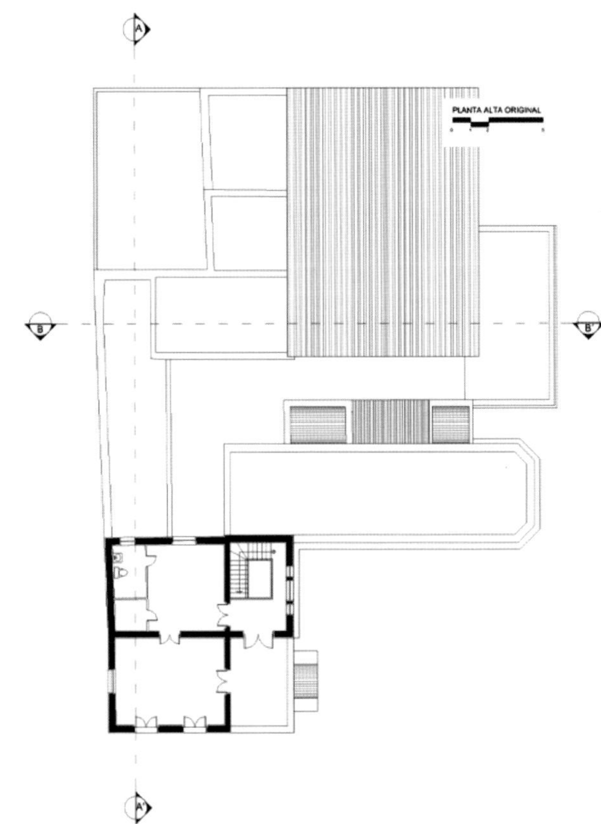

Existing upper level

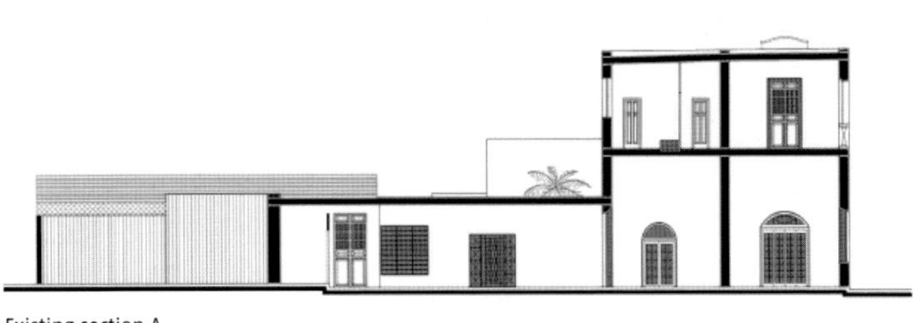

Existing section B

Existing section A

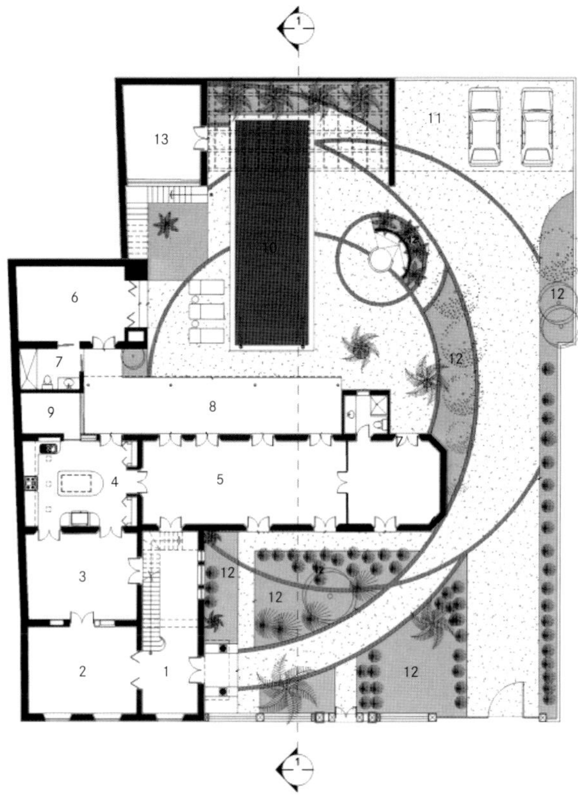

New lower level

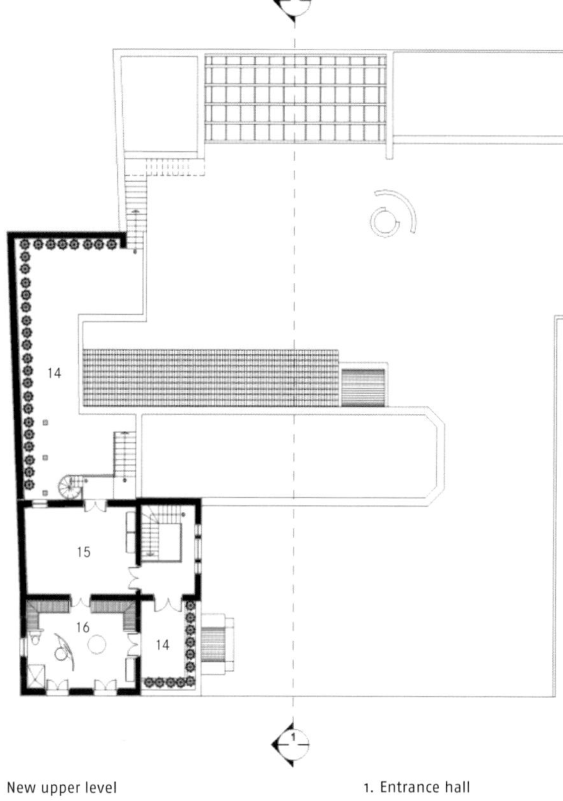

New upper level

1. Entrance hall
2. Studio
3. Dining room
4. Kitchen
5. Living room
6. Guest bedroom
7. Guest bathroom
8. Cover terrace
9. Laundry room
10. Pool
11. Parking
12. Garden
13. Cellar
14. Terrace
15. Master bedroom
16. Dressing room and bathroom

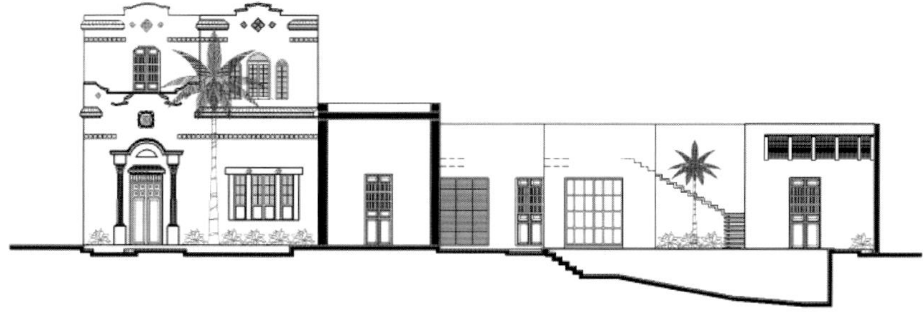

New section

In the original building, the predominance of straight lines set the tone to add new features with curved lines that incorporate smoothness and give the property a contemporary character.

The new kitchen has a strategic location, connecting the common area, the bathroom and the exterior spaces. The design incorporates new and recycled elements; for example, it combines the white concrete of the island with wood and stainless steel in the rest of the kitchen. This mixture creates a space that is functional, contemporary and cozy.

A part of the old factory with sheet metal roof was preserved. When the sun's rays hit the roof, its silhouette is projected onto the pool. This effect, combined with the murals of the artist Michele Maggiora, is very striking.

> Mérida, Yucatán, Mexico | 2007 | Duration of project: 1 year | 2,443 sq ft | © Roberto Cárdenas Cabello <

Estilo Arquitectura

'In cities with a rich architectural heritage, as new architectural expressions are created, it is important that the architecture of the past is rescued and revalued, whether it is colonial, neoclassical, modern, etc.' This premise was the starting point of this restoration process. The new owners' brief was to adapt this everyday family house to a house for resting to be occupied during short spells throughout the year.

Before the remodeling, the house was basically made up of two parts: the house itself and a back yard, which was used as a vegetable garden. The intervention consisted in reusing existing spaces and rearranging them according to the needs of the new inhabitants, as well as maximizing the aesthetic qualities of the materials and forms. All these changes had to maintain the premise of connecting the interior with the exterior.

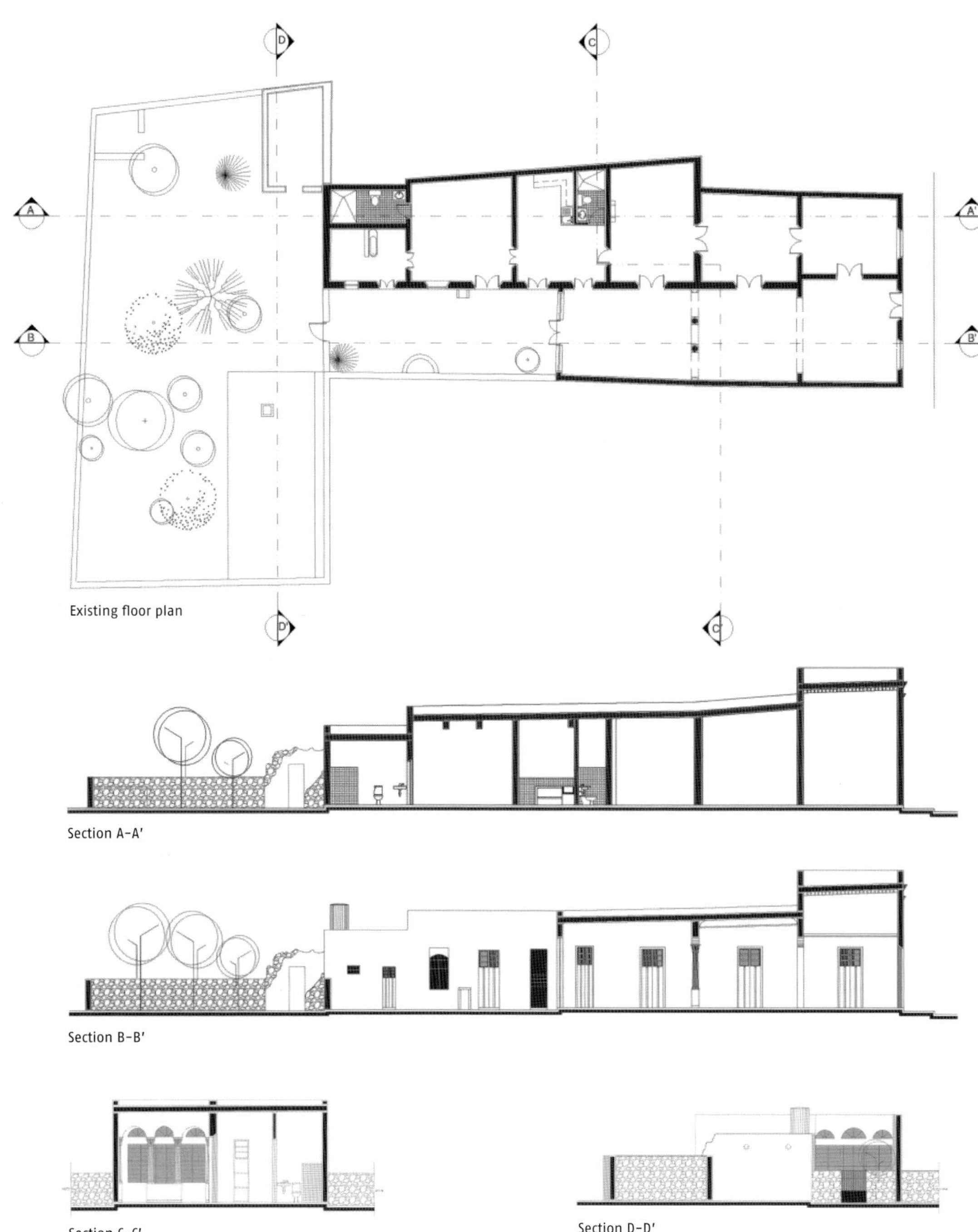

Existing floor plan

Section A–A'

Section B–B'

Section C–C'

Section D–D'

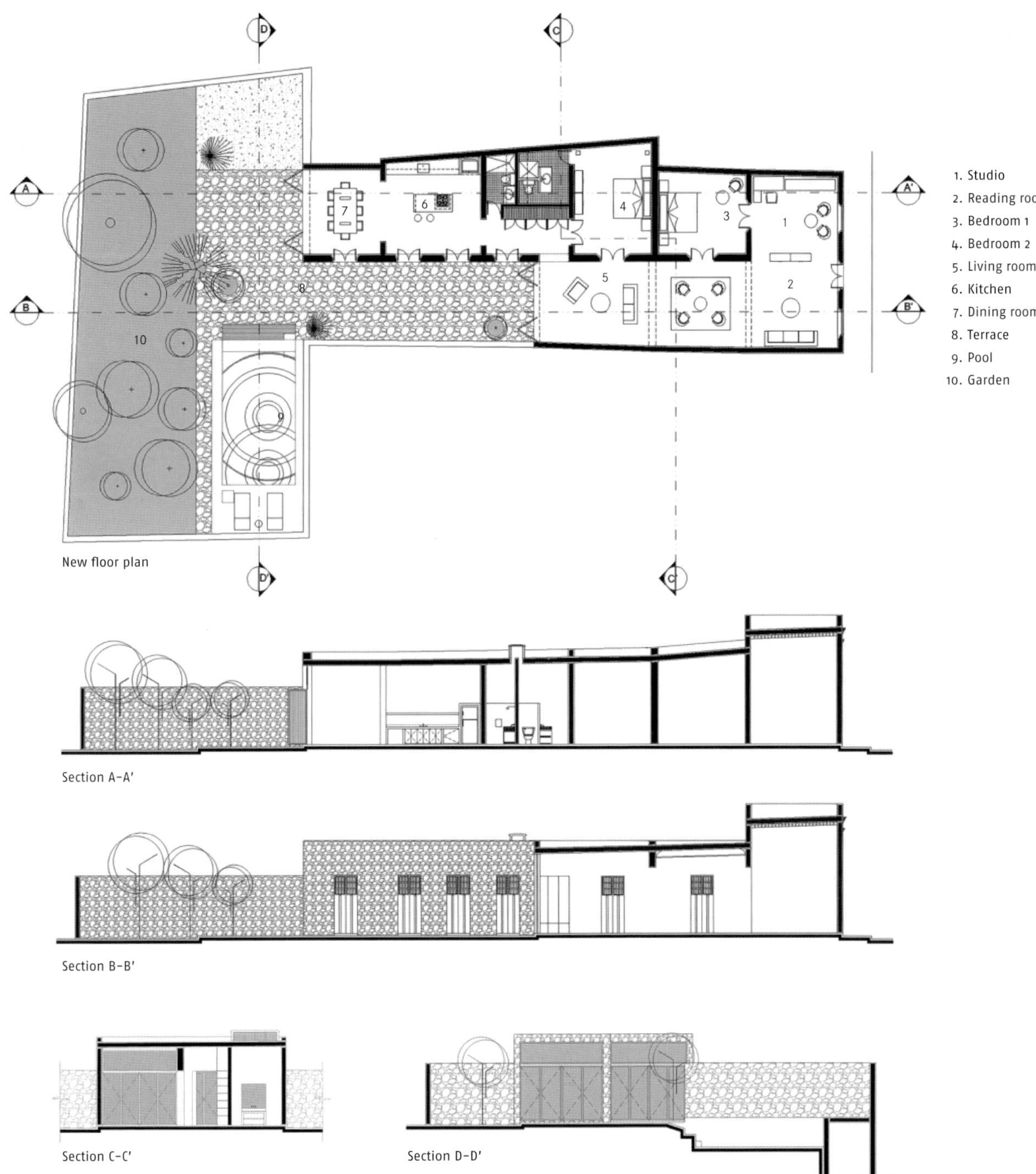

AFTER
New floor plan
Section A-A'
Section B-B'
Section C-C'
Section D-D'
1. Studio
2. Reading room
3. Bedroom 1
4. Bedroom 2
5. Living room
6. Kitchen
7. Dining room
8. Terrace
9. Pool
10. Garden

One of the main challenges of the intervention was to integrate the backyard and vegetable garden into the daily life of the house. The architects also wanted to provide the interior spaces with natural light in order to enhance the relationship with the outside world.

The white, sober interior walls balance the heavy, irregular stone finish of the exterior. The different paving decoration defines the different areas of the building. Neutrality and restraint were the central themes in the layout of the spaces, maintaining previous forms.

The interior spaces were rearranged
to achieve better functionality.
Two parallel blocks were created,
each based on simple forms and
separated by a glazed wall. This
gives light and continuity to the
inner courtyard and garden.

Alla Kazovsky Architects

> Nichols Canyon, CA, USA | 2006 | Duration of project: 8 months | 2,400 sq ft | © Josh Perrin <

Alla Kazovsky, who was both client and architect in this project, wanted a dwelling that would update the property's potential. The combination of new and existing features enhanced the warmth that she felt in the house. Kazovsky created a customized lab, as a space to invent prototypes and to experiment.

The main intervention consisted of completely renovating the house's interior spaces. The kitchen space was opened up, a new dining area and dressing space was created, and the bathrooms were enlarged.

The building's new design and the interior design scheme cleverly fusions the old and the new. Kazovsky conserved the roof, the brick walls, and timber beams, but she laid new white oak flooring. She also designed a new glass and wood front door, which provides the living area with the necessary daylighting.

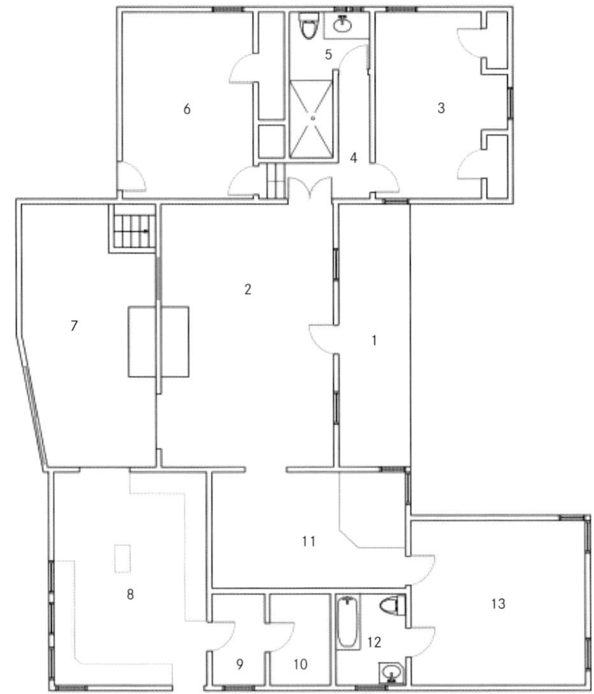

Existing floor plan

1. Porch
2. Living room
3. Bedroom 1
4. Hallway
5. Bathroom 1
6. Bedroom 2
7. Den
8. Kitchen
9. Laundry
10. Pantry
11. Bedroom 3
12. Bathroom 2
13. Master bedroom

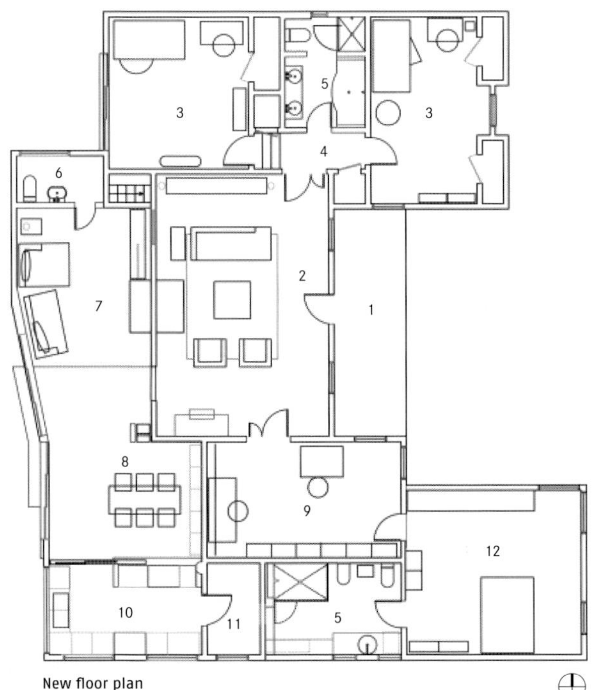

New floor plan

1. Porch
2. Living room
3. Bedroom
4. Vestibule
5. Bathroom
6. Powder room
7. Family room
8. Dining room
9. Office
10. Kitchen
11. Laundry
12. Master bedroom

Floor plans before and after the remodeling. One of the main aspects of the renovation was reducing the kitchen to free up space for a new dining area. The space between the lounge and the master bedroom was also remodeled into a work space.

The architect sought different solutions to separate the new remodeled spaces.
One solution consisted of big sliding glass dividers that were used to connect the dining room and a part of the living room with the patio.
A translucent glass sliding door was also installed between the kitchen and the living room.

A new architects' studio was installed in the former garage. The traditional blinds were replaced by folding glass doors. The central post of the structure was removed and a horizontal iron beam was inserted. Main features include a bespoke MD wood unit lined with aluminum and a meeting table on wheels.

STONEHEDGE RESIDENCE

> West Lake Hills, TX, USA | 2004 | Duration of project: 3 years | 2,041 sq ft | © Paul Finkel <

Miró Rivera Architects

The main challenge in this project was to work in a house that had recently been renovated but where the owners still felt the need to make finishing touches.

The architects' main objective was to improve the house's connection with the pool and garden, and to amplify its interior common spaces. Three important actions were carried out to achieve these objectives. The first focused on remodeling the entrance, which was reconfigured to clearly define its use. In the dining room, a hall with a copper clad pivot door was created, which welcomes visitors. The second action focused on the master bedroom, which was enlarged, allowing a better hierarchy of spaces. Finally, the terrace and trellis were created to unify the house and garden.

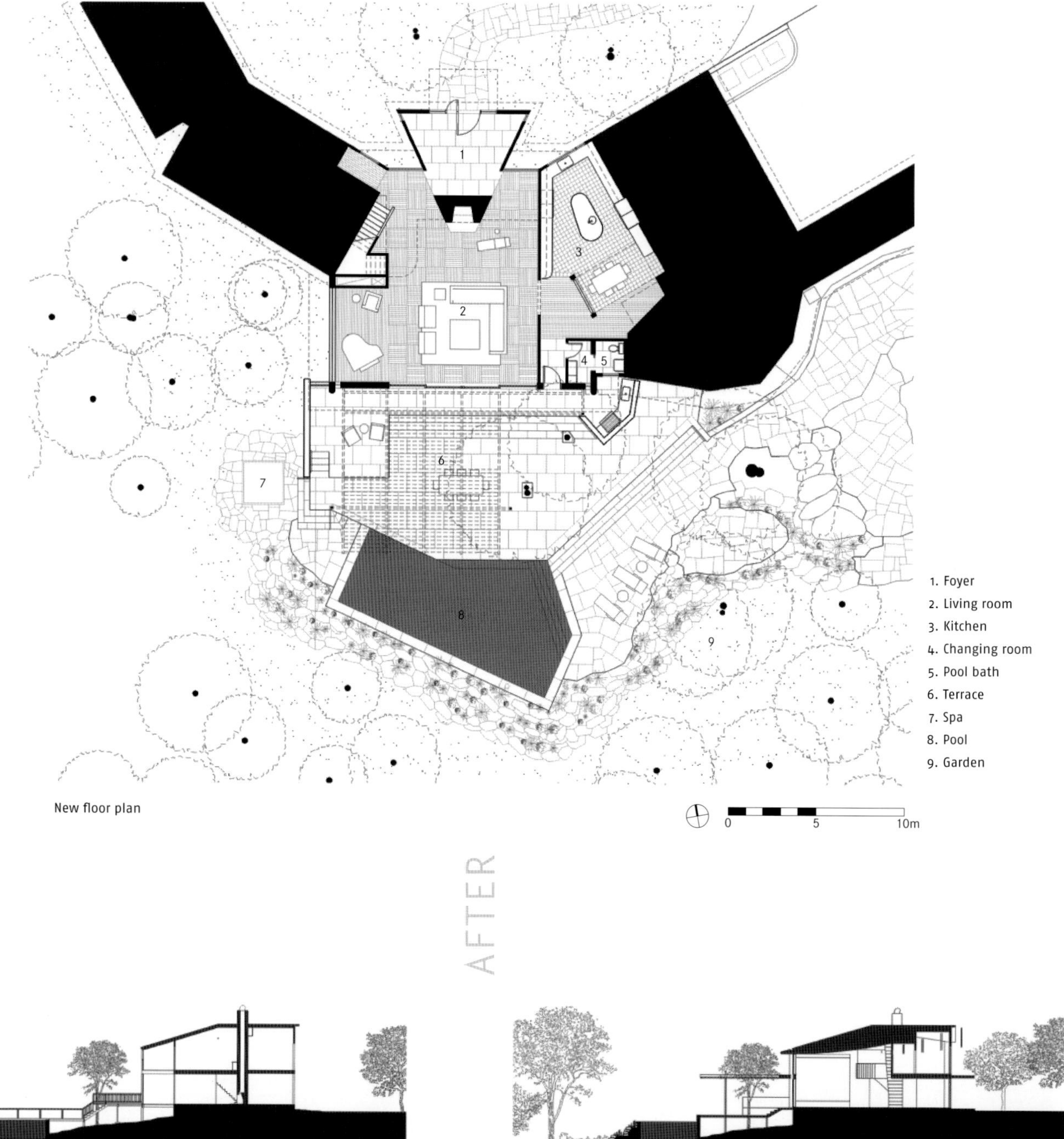

1. Foyer
2. Living room
3. Kitchen
4. Changing room
5. Pool bath
6. Terrace
7. Spa
8. Pool
9. Garden

New floor plan

0   5   10m

BEFORE

AFTER

Existing section

New section

0   5   10m

A natural stone path was created to integrate the xeriscape into the new entrance. The architects designed big windows, a small porch and a swing door to access the hall.

Saia Barbarese Topouzanov Architectes

> Montreal, Canada | 2007 | Duration of project: 8 months | 3,100 sq ft | © Vladimir Topouzanov <

This centennial home near Montreal, was transformed by the architects into a practical and luminous space. The remodeling focused on increasing daylighting and ventilation through opening the house out onto the garden. This created large and flexible interior spaces.

The main work was carried out on the roof coverings, which limited the amount of light that was able to enter. The doors and windows at the rear of the house were enlarged to increase communication with the patio. Materials that had been damaged by the passage of time were replaced, and the original façades were cleaned. The general structure of the house was renovated in a way that conserved the building's original form. The façade design was inspired by the *Campsis radicans* in the patio through the use of colored bricks.

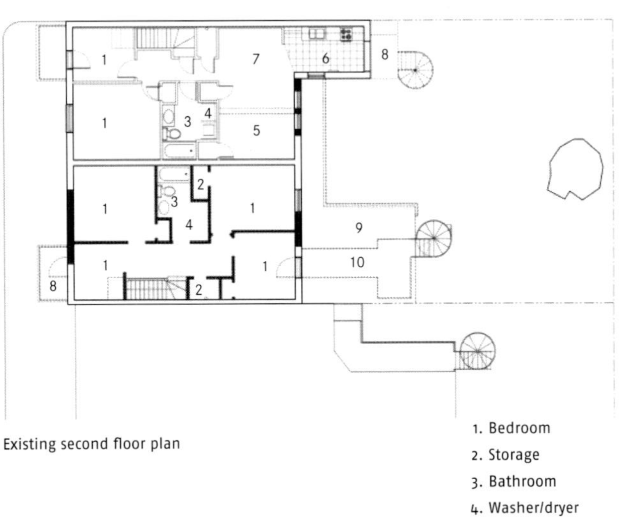

Existing second floor plan

1. Bedroom
2. Storage
3. Bathroom
4. Washer/dryer
5. Living room
6. Kitchen
7. Dining room
8. Terrace
9. Roof
10. Bridge

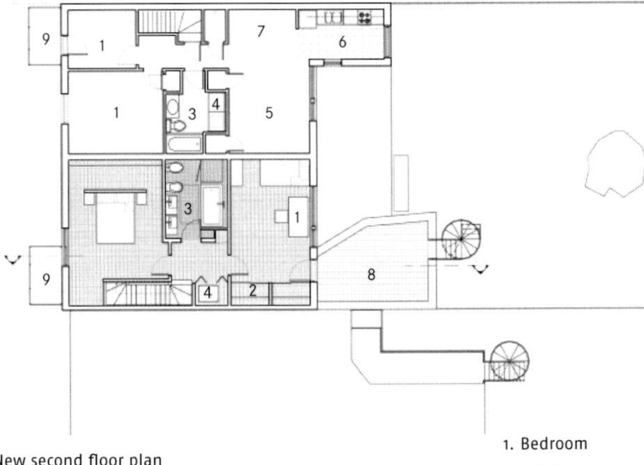

New second floor plan

1. Bedroom
2. Closet
3. Bathroom
4. Washer/dryer
5. Living room
6. Kitchen
7. Dining room
8. Terrace
9. Balcony

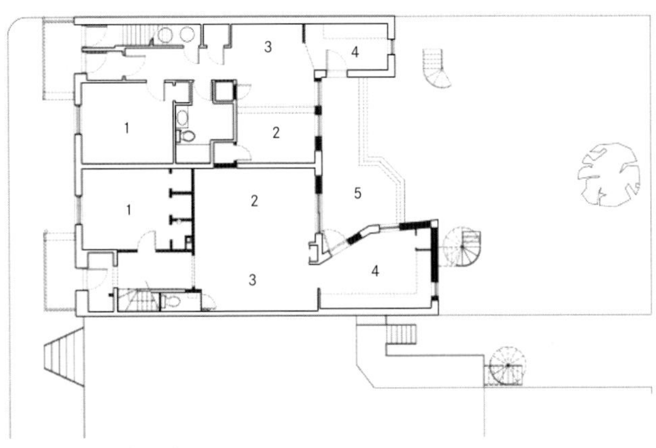

Existing ground floor plan

1. Bedroom
2. Living room
3. Dining room
4. Kitchen
5. Terrace

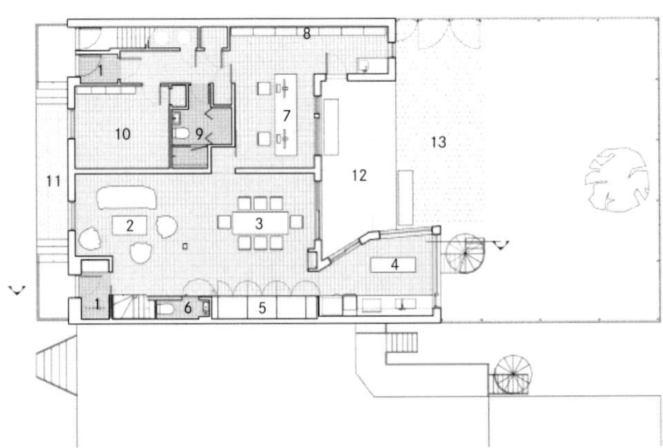

New ground floor plan

1. Entry
2. Living room
3. Dining room
4. Kitchen
5. Storage
6. Bathroom 1
7. Studio
8. Library
9. Bathroom 2
10. Bedroom
11. Entrance pavement
12. Terrace
13. Reinforced grass

0   5m

Floral wallpaper inspired by the trumpet vine of the courtyard

Unfolded elevation

Street elevation

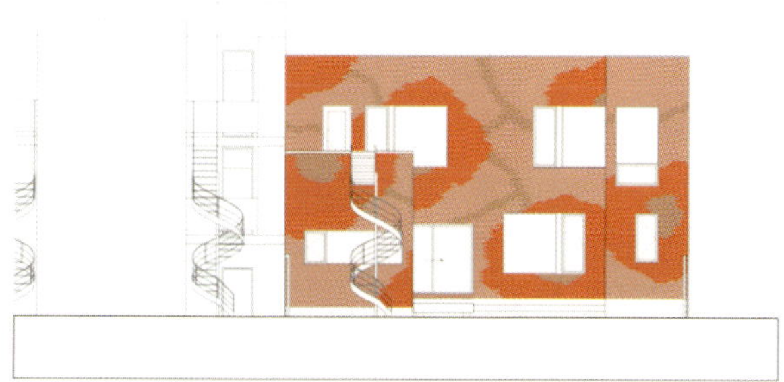

Garden elevation

Side elevation

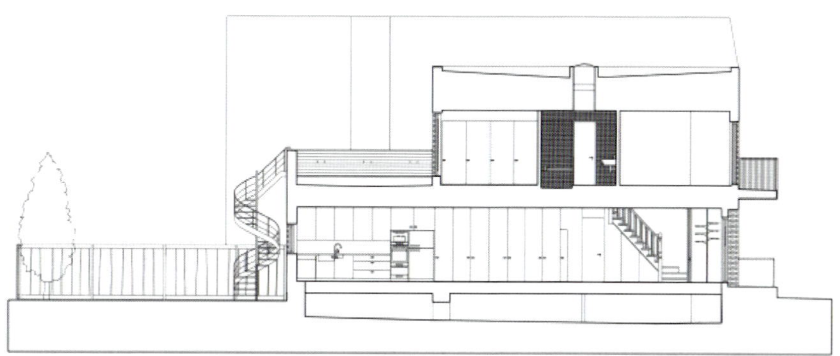

Longitudinal section

The alterations made in this house divided the spaces into two floors: the common spaces on the first floor, and the private ones on the second. A roof terrace was created in place of the original slanted roof. The final remodeling phase focused on the outdoor spaces. To minimize maintenance, perforated concrete flooring was used for the street access, while the garden transit areas were reinforced with a transparent plastic fitted below the grass, creating a durable surface. A spiral staircase connects the master bedroom to the garden.

The mobile kitchen island enabled the space to be enlarged and allowed the use of independent hook-ups. The walls, cupboard doors and surfaces are finished in bright white to emphasize the abundant lighting. This solution contrasts with the original staircase, whose structure has been preserved intact.

In the living room, a long room that extends to the back garden was created by knocking through some internal walls. This open space offers views of both sides of the lot and allows daylighting during most of the day. To make the space are flexible as possible, one wall has been completely transformed to accommodate a staircase leading to the first floor, an iron cupboard, a washroom, a wardrobe, a small storage area, a wine cellar, a pantry, and several kitchen appliances.

CONNOR RESIDENCE

> Elwood, VIC, Australia | 2009 | Duration of project: 5 months | 1,292 sq ft | © Jesse Marlow <

The central design concept of the Connor house was to create a series of new features inserted into the existing space. These would not compete with the apartment's original art deco design, but enhance it. The features added consist of modern wood designs that are carefully elaborated to instill an elegant air that blends simplicity and luxury.

On the other hand, the interior spaces in the house have been designed to create a cozy atmosphere. The client, a film producer, helped to define the volumes, shapes and compositions. The furniture is an active component in the house; with a little imagination these compositions become faces with mobile panels that look like mouths. There are also elongated horizontal elements that evoke arms and wings. These simple analogies make the occupants feel at home.

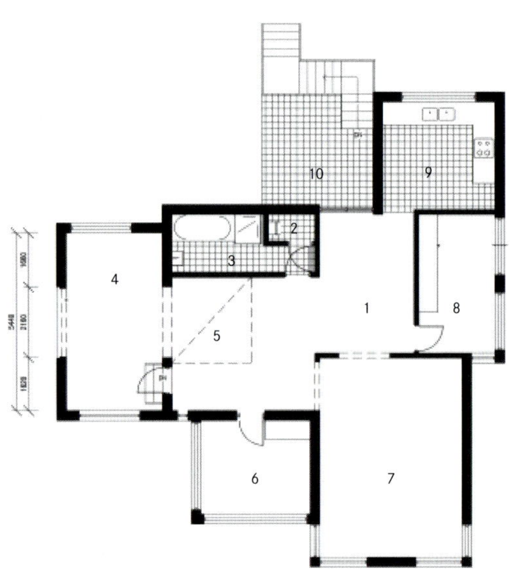

Existing floor plan

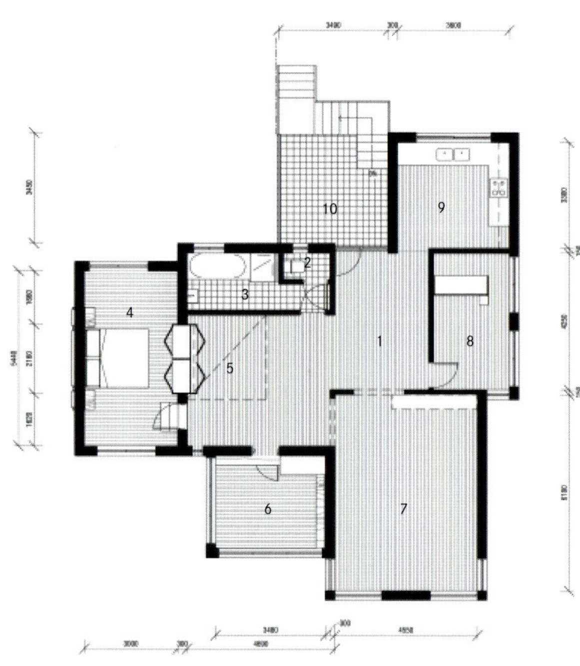

New floor plan

1. Entry
2. Bathroom
3. Master bathroom
4. Bedroom 1
5. Dining room
6. Bedroom 3
7. Living room
8. Bedroom 2
9. Kitchen
10. Terrace

1. Entry
2. Bathroom
3. Master bathroom
4. Bedroom 1
5. Dining room
6. Bedroom 3
7. Living room
8. Bedroom 2
9. Kitchen
10. Terrace

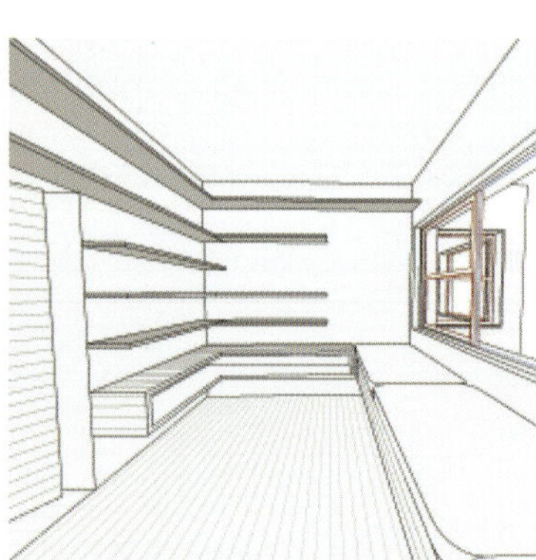

Computer generated interior views

A variety of wood elements are arranged throughout the apartment to breathe life into the space through their warm shades and simple yet emphatic volumes. Each piece of wood furniture −whether it is a cupboard, minibar, shelves, headrest, sofa−bed, etc− awakens curiosity in the onlooker.

Recycled wood was used to create furniture. Two varieties of eucalyptus were chosen. The planks were arranged according to their different sizes to define the pieces of furniture and therefore prevent unnecessary joints that would deconstruct the overall appearance of each piece.

Recycled wood elaborated with a simple oil finish combined with careful joinery guarantee the pieces will be longer lasting. Some of the pieces lack fixtures, giving them a singular aspect of being structures without strange elements that could drain the power of each piece.

Carola Vannini Architecture

> Roma, Italy | 2007 | Duration of project: 8 months | 1,292 sq ft | © Filippo Vinardi <

The main objective of the architects was to create a new space that met the client's brief to completely change the distribution and decoration of interior spaces.

The main intervention was carried out in the common areas, where the space has been opened up to create a multitude of perspectives. In line with this premise, the kitchen has been partly concealed behind a green brick wall that does not reach the roof, thus reinforcing the idea of a flowing space.

The rest areas have been designed to be completely private: a sliding door gives access from the common areas to the master bedroom, the office, guest room and the two bathrooms. The master bathroom uses the colors black, white and red to create a warm, relaxing space.

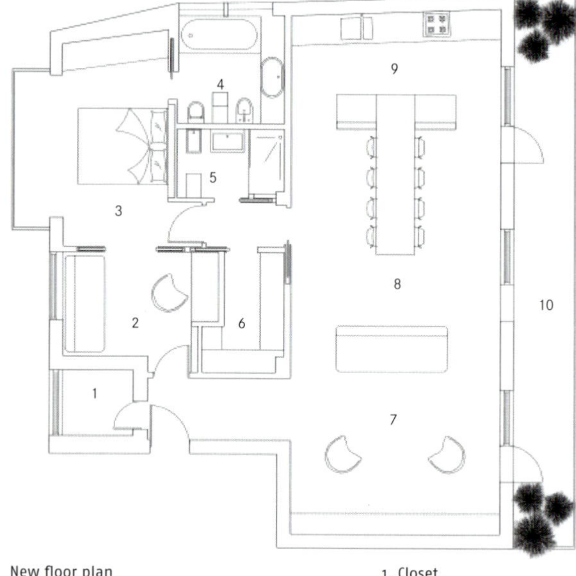

Existing floor plan

1. Guest bedroom
2. Kitchen
3. Bathroom
4. Master bathroom
5. Master bedroom
6. Living room
7. Balcony

New floor plan

1. Closet
2. Guest bedroom
3. Master bedroom
4. Master bathroom
5. Bathroom
6. Walk in closet
7. Living room
8. Dining room
9. Kitchen
10. Balcony

The kitchen has been designed to be partly separated from the living room. This was achieved by placing a green brick wall designed as if it were a curtain. An opening in the center makes space for a long table that protrudes beyond the wall. This structure serves as a dining table in the living room and as a shelf in the kitchen.

In the remodeling, colors and materials have taken on an important role: the mix of bottle green (in the brick wall, the sofa, and the kitchen chairs), brown (on the flooring and furniture) and white (on the walls) infuses the space with a feeling of well-being and relaxation. These colors become the unifying feature throughout the house.

The master bedroom and guest room are separated by a large sliding door. This allows the owners to either use the small room as office space or to integrate it into the master bedroom. Whatever option is chosen, the concept of continuous space prevails in this remodeled house.

Carola Vannini Architecture

> Roma, Italy | 2007 | Duration of project: 6 months | 1,292 sq ft | © Filippo Vinardi <

The renovation of this private residence located in the Trastevere area in the heart of the old city of Rome included completely reorganizing its interior spaces and designing new furniture and lighting systems.

The project's main objective was to balance the older, rustic outer structure of the building with its modern-style interiors. The common areas were designed in an open-plan program that interacts with the large garden. The kitchen is partly concealed behind a brick wall that does not reach the ceiling, so as to preserve a design scheme of flowing space.

In the new layout, the private areas are divided into two bedrooms, three bathrooms and two built-in wardrobes. In addition to defining the spatial renovation, the architects also designed the bespoke furniture. The bold forms of the interior design scheme are repeated in the bathrooms through the use of resins and new materials.

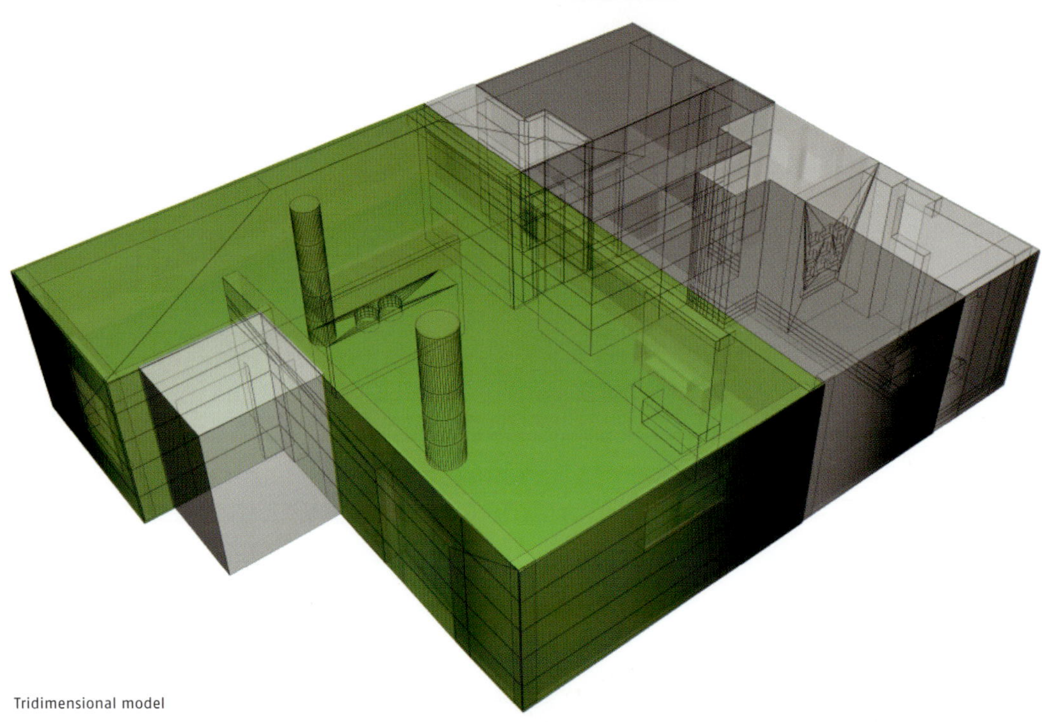

Tridimensional model

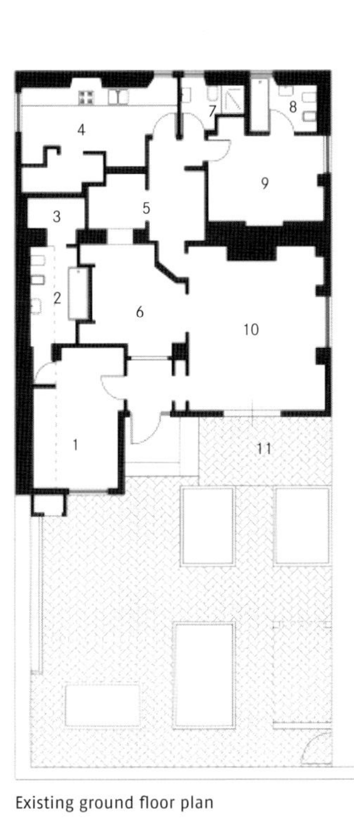

1. Bedroom 1
2. Bathroom 1
3. Walk in closet
4. Kitchen
5. Office
6. Dining room
7. Bathroom 2
8. Bathroom 3
9. Bedroom 2
10. Living room
11. Garden

Existing ground floor plan

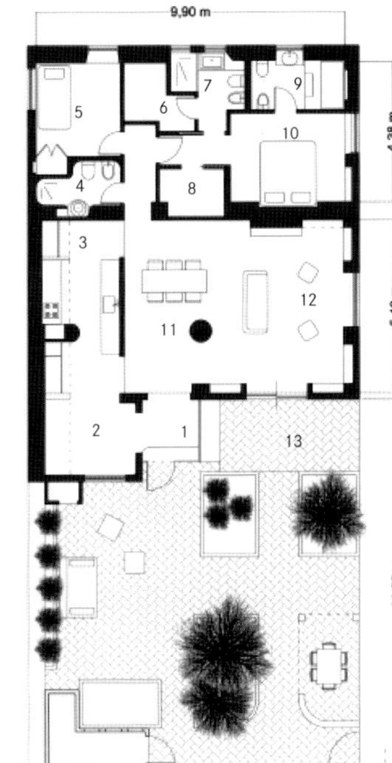
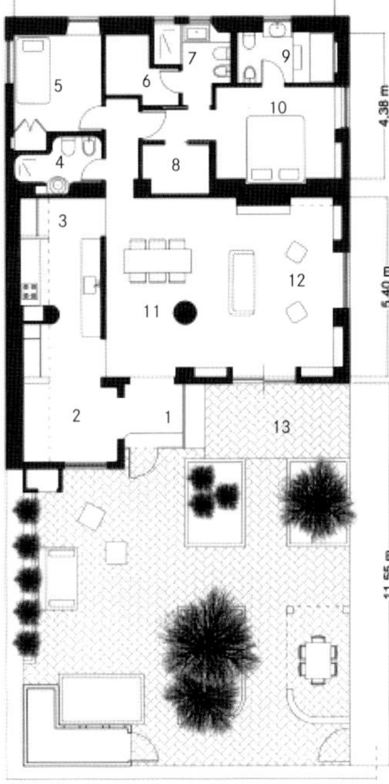

1. Veranda
2. Office
3. Kitchen
4. Bathroom 1
5. Bedroom 2
6. Walk in closet 1
7. Bathroom 2
8. Walk in closet 2
9. Bathroom 3
10. Bedroom 1
11. Dining room
12. Living room
13. Garden

New ground floor plan

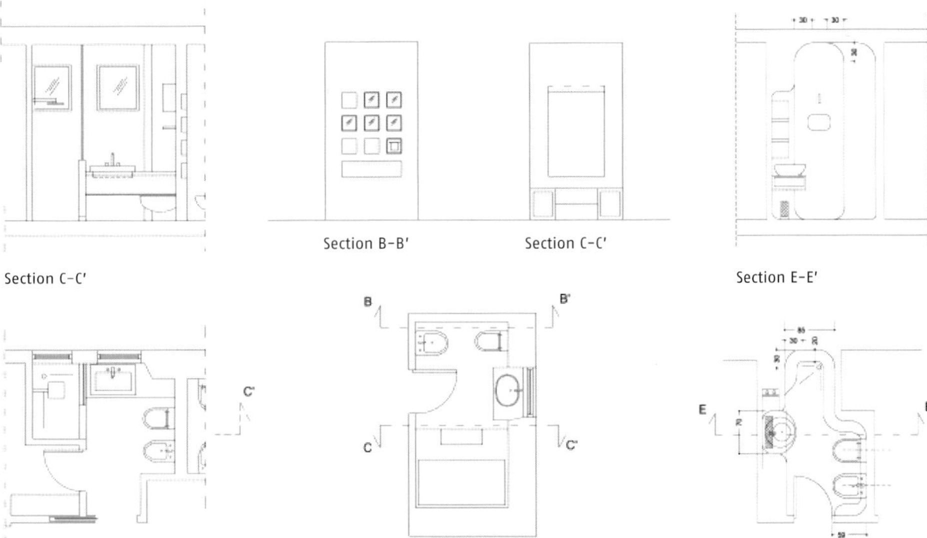

Photographs of the living room and fireplace taken during the remodel. Walls that contained large bookshelves have become monochromatic, smooth surfaces to visually emphasize the fireplace. Former Baroque elements have been removed to make way for a minimalist space where colors take center stage.

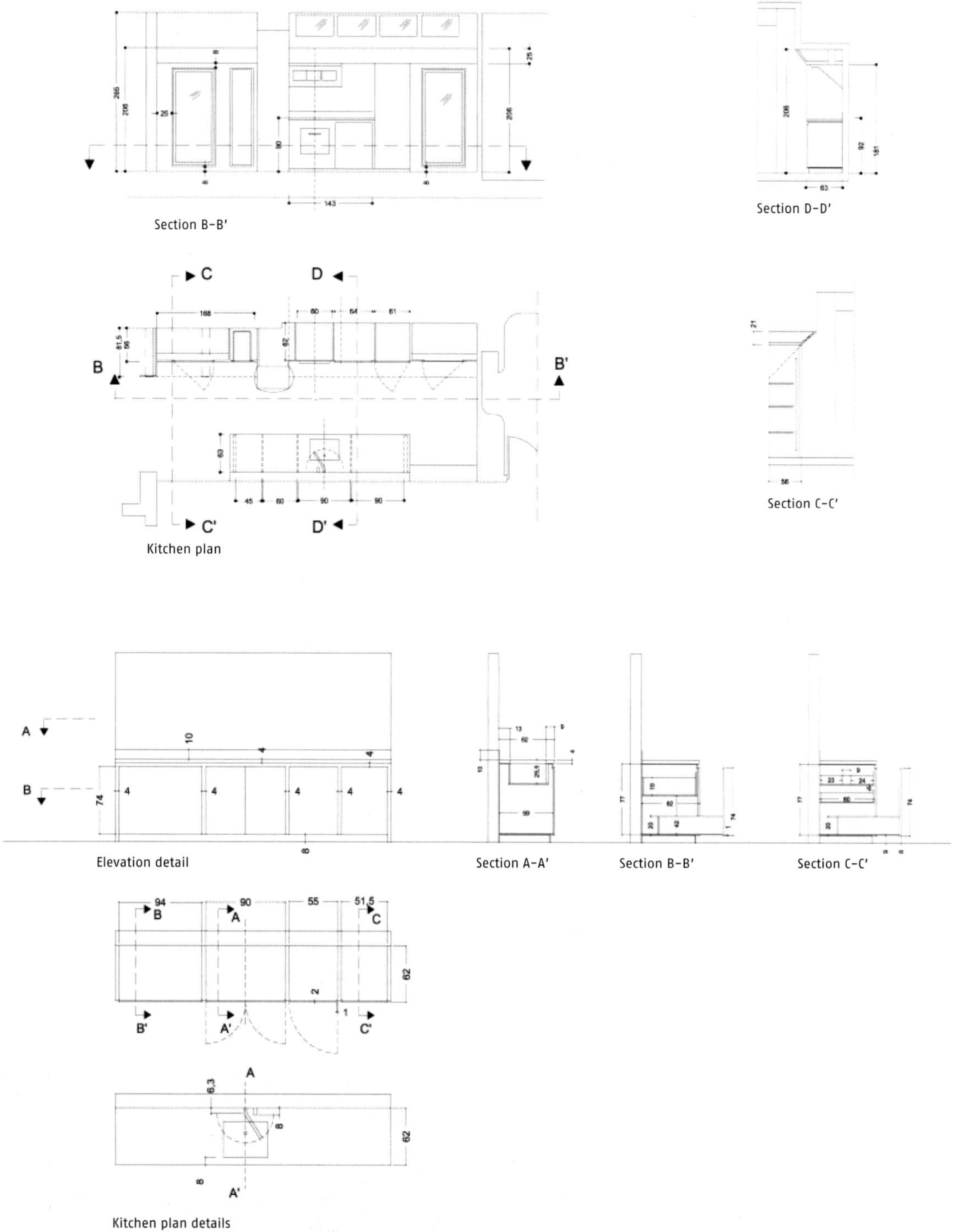
Section B-B'
Section D-D'
C
D
B
B'
C'
D'
Kitchen plan
Section C-C'
A
B
Elevation detail
Section A-A'
Section B-B'
Section C-C'
B
A
C
B'
A'
C'
A
A'
Kitchen plan details

The architects placed special emphasis on making the modern fireplace, which retains traditional elements such the brickwork, visible from both the entrance of the house and the garden. If it were a perspective painting, the fireplace would be the vanishing point.

As the architects wanted to create an open space, most of the kitchen is visible from the living room. A minimalist style was chosen to emphasize visual continuity. Light colored oak wood was used to harmonize the design scheme. At one end of the space there is an opaque glass door fitted with a green neon light. This decorative element connects with the same color tones used in the living room sofa.

The small, irregular surfaces in the three bathrooms have been transformed and all the angular details have been eliminated to eradicate any visual reference point. The white bathroom has been designed to create a surreal feeling by using shiny resins and wood furnishings with glossy enamel. The black bathroom features a combination of dark tones on the walls and in the furniture, which contrast with the white bathroom suite. Finally, the red bathroom has been designed as a geometric space with a minimalist aesthetic. Here, the color red is interrupted by the wooden furniture with glossy white enamel.

Stelle Architects

> Bridgehampton, NY, USA | 2008 | Duration of project: 2 years | 3,300 sq ft | © Jeff Heatley <

Architecture and nature have been fully integrated in this project, which includes a guest house, a garage for two cars, a pool and a two-story house. The building is located on the coast, a factor that the architects took advantage of to create a structure with big windows that provide both daylighting and fantastic views.

The original building, which had become run down over the years, is raised on stilts above the dunes. The house was emptied and extended in accordance with local building regulations. The intervention focused on the original structure, which was reinforced and framed with steel, and in the new layout, and the exterior aesthetics. The dwelling is clad with wood and cement board forming a shield protecting against the rain. Anodized aluminum structures were used to frame the windows.

GLASS
FRAGILE
HANDLE
WITH CARE

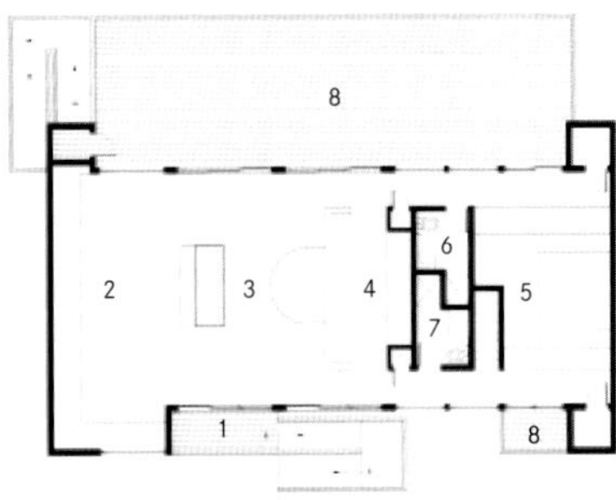

Existing second floor plan

1. Entry
2. Living room
3. Dining room
4. Kitchen
5. Master bedroom
6. Bathroom 1
7. Bathroom 2
8. Deck

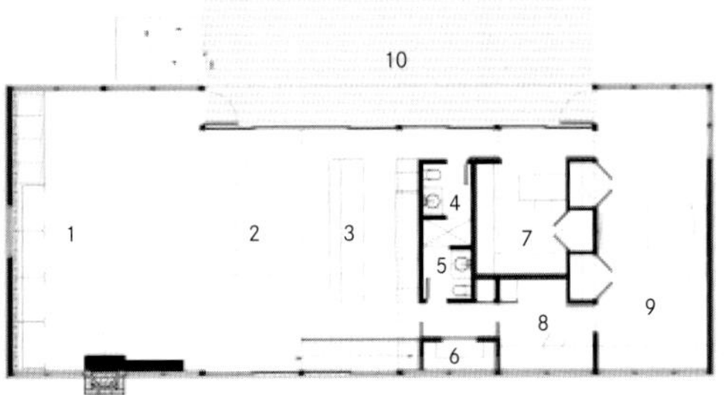

New second floor plan

1. Living room
2. Dining room
3. Kitchen
4. Bathroom 1
5. Bathroom 2
6. Pantry
7. Office 1
8. Office 2
9. Master bedroom
10. Deck

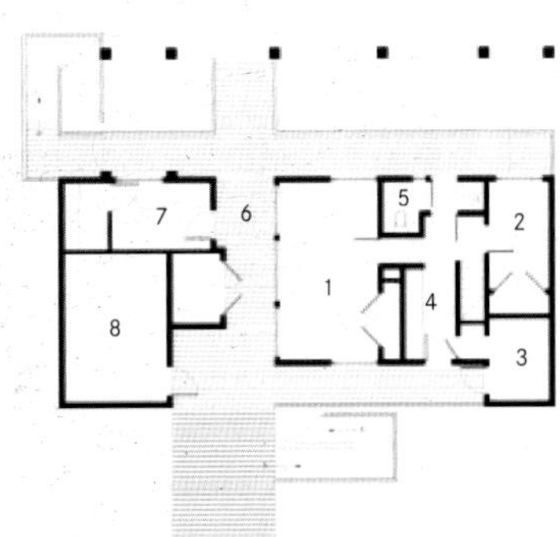

Existing ground floor plan

1. Bedroom 1
2. Bedroom 2
3. Laundry
4. Vestibule
5. Bathroom
6. Boardwalk
7. Sauna
8. Mechanical room

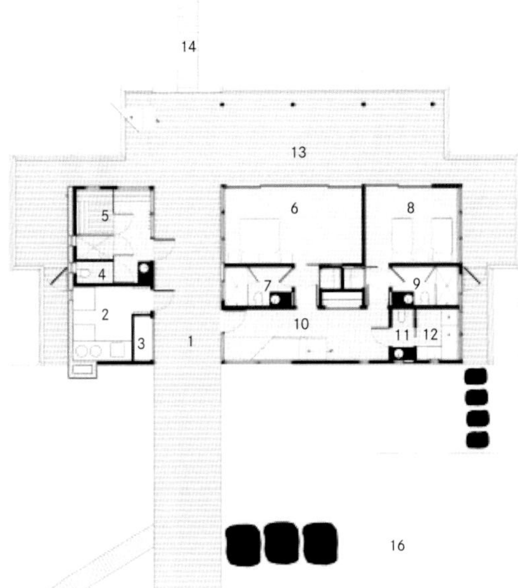

1. Breezeway
2. Mechanical room
3. Trash
4. Bathroom 1
5. Sauna
6. Bedroom 1
7. Bathroom 2
8. Bedroom 2
9. Bathroom 3
10. Entry vestibule
11. Powder room
12. Laundry
13. Covered deck
14. Boardwalk 1
15. Boardwalk 2
16. Parking

New ground floor plan

0   5   8m

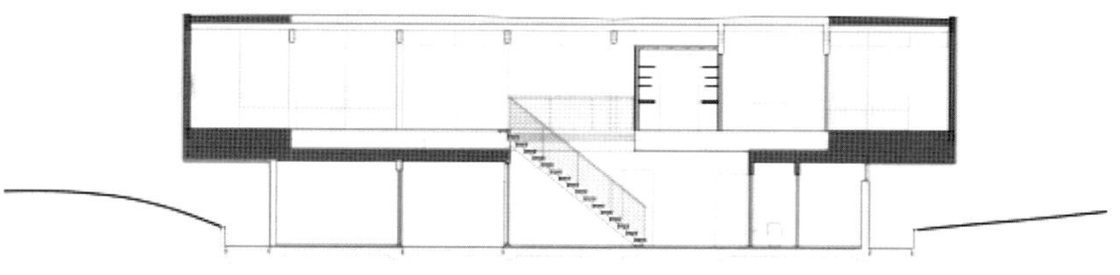

Longitudinal section

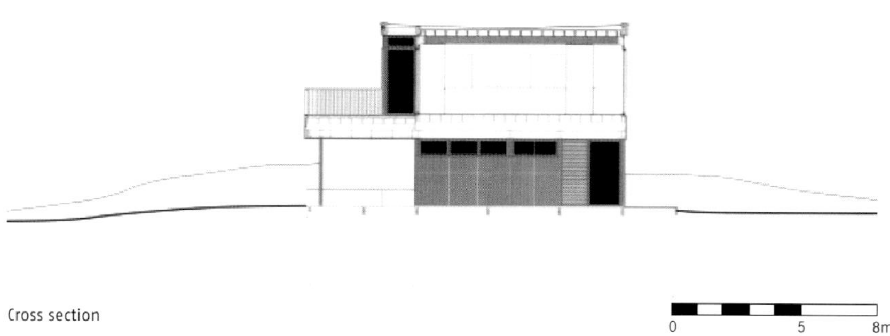

Cross section

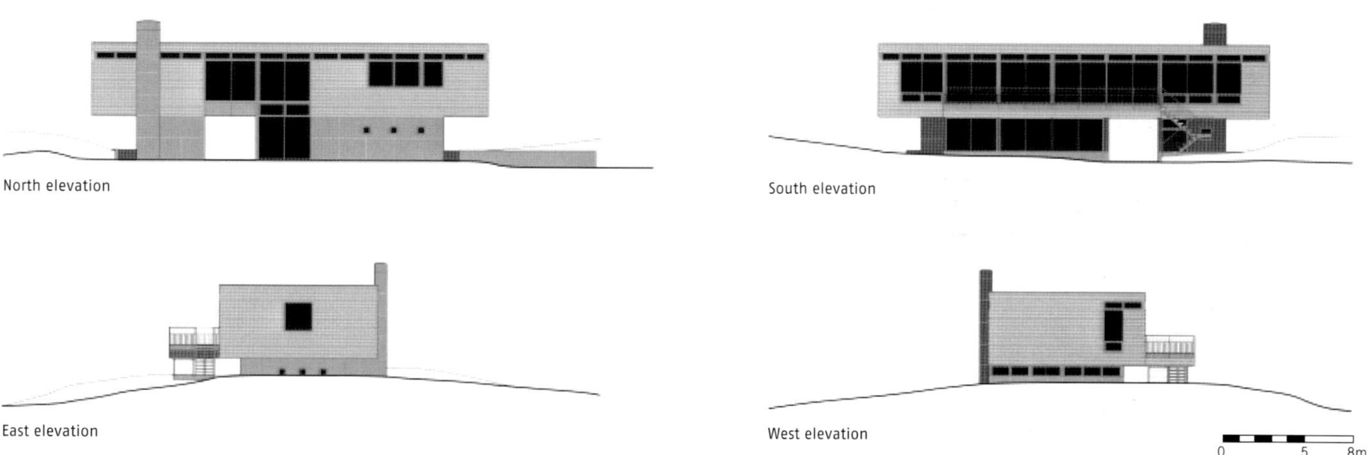

North elevation

South elevation

East elevation

West elevation

During the renovation process, the original house was emptied and extended to meet the client's needs. The small windows, which made the wooden interiors gloomy, were replaced by large windows with anodized aluminum frames, providing light and color.

The architects designed a sauna that is entered through the corridor. The new house seems to float elegantly over the dunes, while maximizing the ocean views. The natural area surrounding the house was replenished with native vegetation.

The second floor is designed as if it were a loft: a kitchen, lounge and dining room are located in a single elongated space. The doors maximize cross-ventilation. Both the guest house and the main dwelling are equipped with geothermal air conditioning, as well as photovoltaic panels.

Driendl Architects

> Vienna, Austria | 2003 | Duration of project: 13 months | 1,292 sq ft | © Driendl Architects <

This early 20th century house was remodeled to adapt to the client's family living requirements. The property is located in district 19 of the Austrian capital, in an area of buildings constructed in a similar manner. The house had several anomalies, such as its narrow space, terrible lighting conditions and zero relationship between the building and the garden.

With these characteristics, the architects decided to open up one end of the house to obtain more daylighting and space. The building lies in a long, narrow, sloping lot that faces north.

The most important intervention was adding a south-facing annex with a spacious terrace and garden views. The house's old southern façade was totally demolished and the structure was stabilized by means of a steel frame. In front of this open façade, a new 283 ft³ cube was erected.

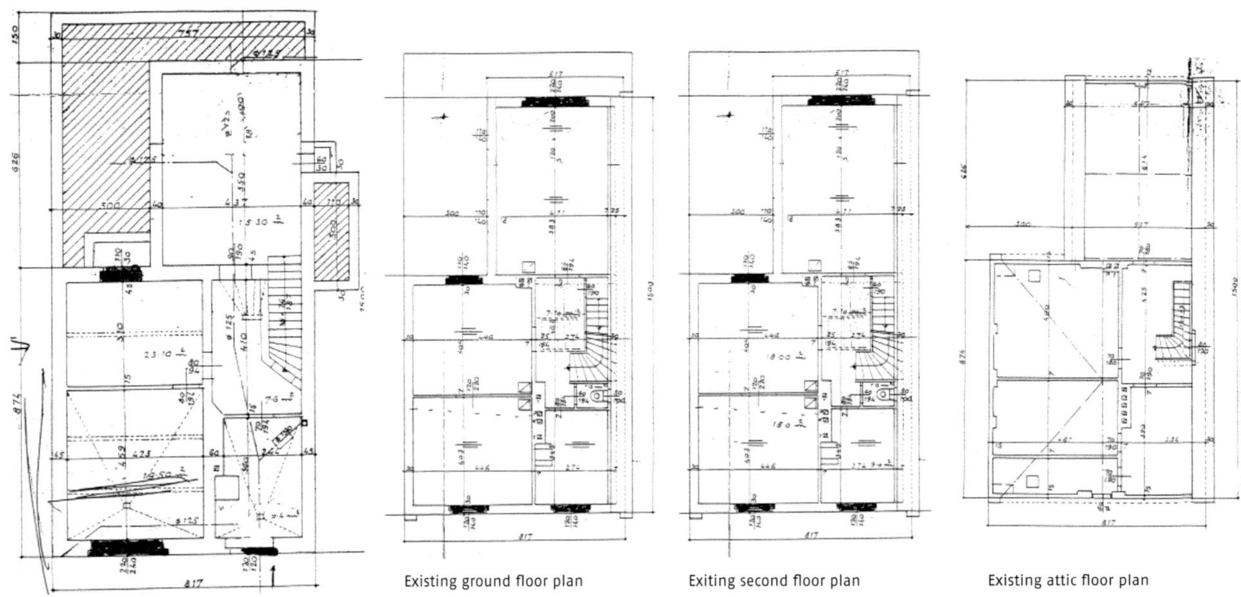

Existing basement floor plan

Exiting second floor plan

Existing ground floor plan

Existing attic floor plan

Existing front elevation

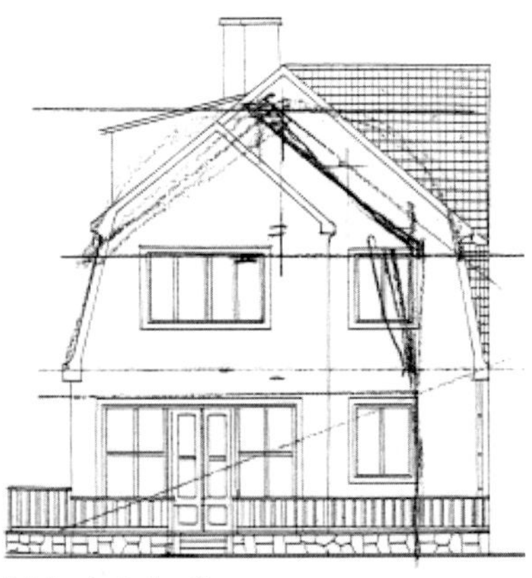

Existing back elevation

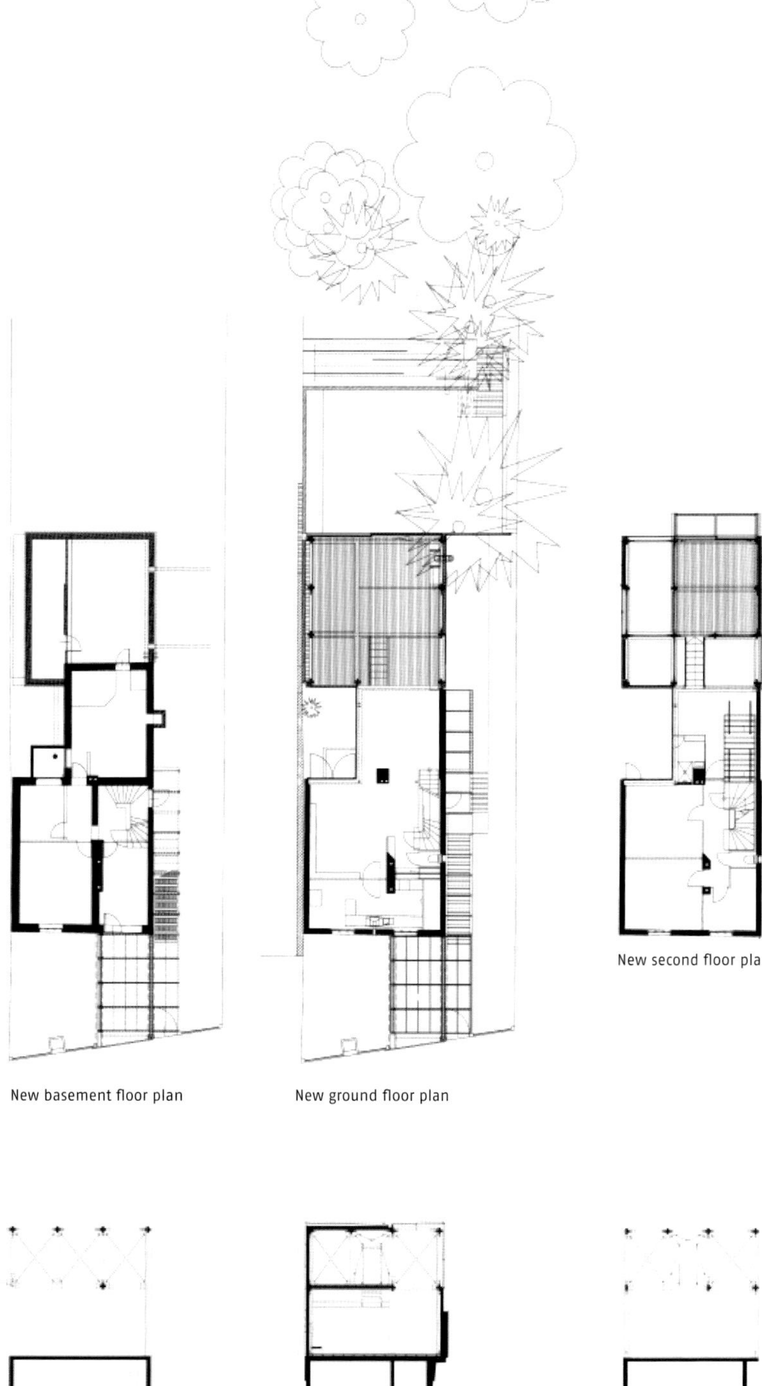

New basement floor plan

New ground floor plan

New second floor plan

New sections

The architect's most important intervention was building an annex that provides the building with more space and daylighting, illuminating the rooms located in the old part of the house. An atrium was also created; this is another example of the harmonious relationship between the old, the new and nature.

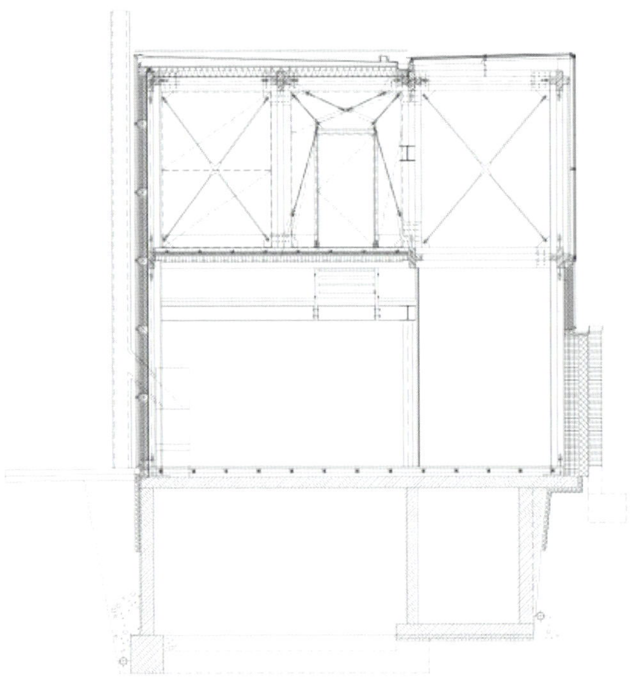

New detailed section

Steel frame suspension

Due to its open structure and width, the proportions of this cube offer panoramic views of the garden and its big windows wash the space with daylighting. The architect opted to construct the cube using a steel and wood structure. The lower floor of the cube is large living space, without pillars, meaning the construction features do not interrupt the views. On the floor above there is a bedroom with balcony. The cube also has a roof terrace.

Architects EAT

> Hawthorn, VIC, Australia | 2007 | Duration of project: 16 months | 1,938 sq ft | © Craig Shell, Rhiannon Slatter <

The philosophy and methodology applied by the architects focused on using natural light and shade as part of the architecture. To achieve this effect they used materials such as aluminum and glass, which dominate in the rear façade.

The structure is a lattice of horizontal glass and aluminum louvers. These allow natural light to enter during most of the day, while creating a light and shade effect. This modular structure is regulated according to the clients' needs. The interior design therefore changes according to the time of day and the owners' wishes.

This remodeling was designed using sustainable criteria that provide well-being thanks to good cross-ventilation, energy savings through daylighting, and a non-polluting thermal heating system.

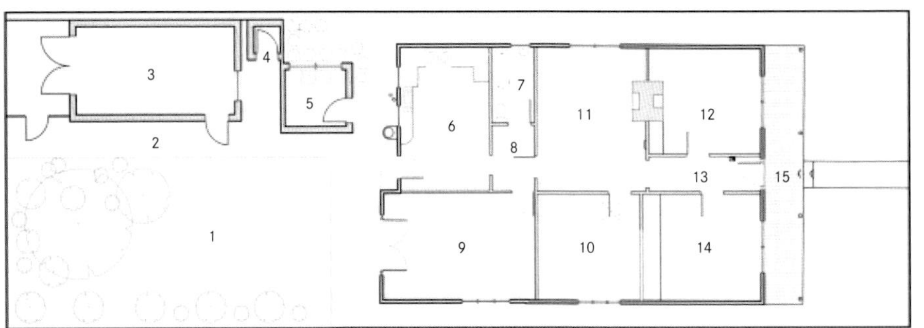

Existing floor plan

1. Vegetation
2. Paver
3. Garage
4. Bathroom 1
5. Laundry room
6. Kitchen
7. Bathroom 2
8. Hallway
9. Dining room
10. Studio
11. Living room
12. Bedroom 2
13. Entry
14. Bedroom 1
15. Covered verandah

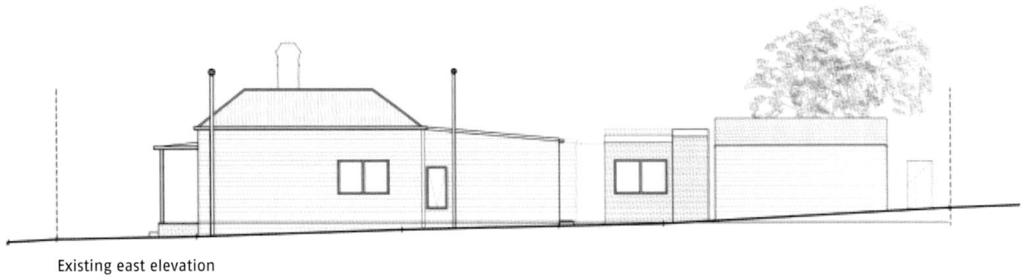

Existing east elevation

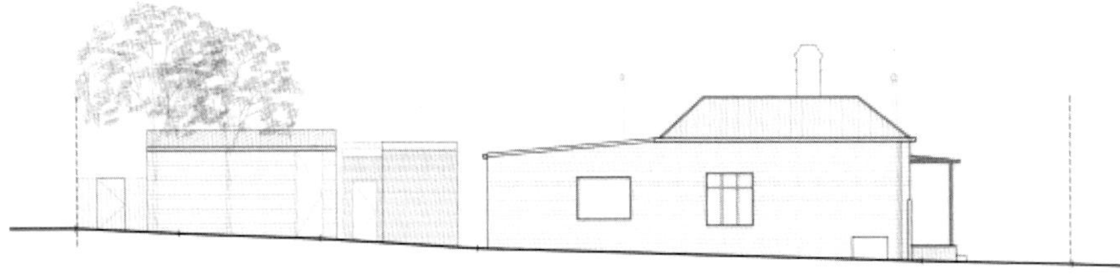

Existing west elevation

Existing north elevation

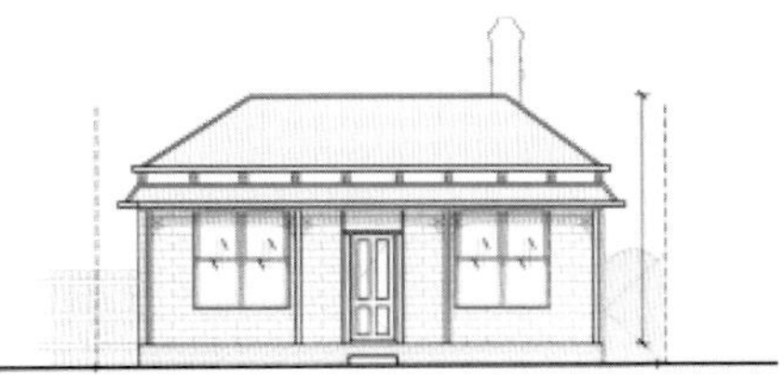

Existing south elevation

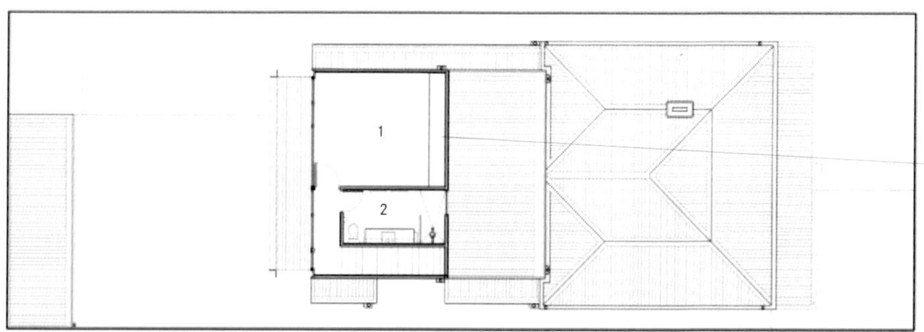

New second floor plan

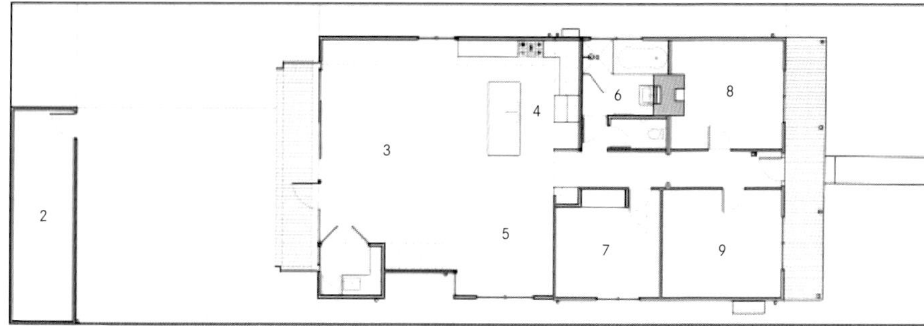

New ground floor plan

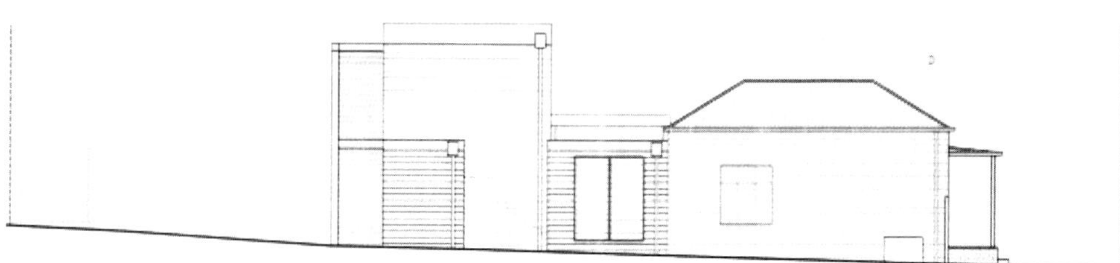

West elevation

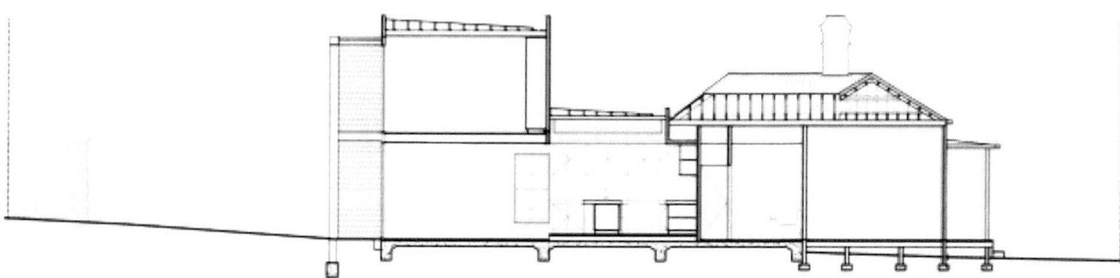

Longitudinal section

The external, fixed aluminum slats were a great resource conceived by the architects for this project. They were used in the north–facing façade as an effective means of shading, and as insulation screens to create privacy. This structure, combined with the shutter-style horizontal glass windows, creates stunning effects of light and shade in the interior, which change depending on the amount of sunlight. Artificial lighting has been installed at strategic points to enhance the structure.

The house was painted bright white to further intensify the light and shade effect. The furniture in the dining room and kitchen is also white to emphasize the architects' intention. All these features highlight the linear nature of the design. The only contrast generated at the two levels is the dark parquet flooring.

In this dwelling, lighting is conceived as another architectural feature. The incidence of natural light on the white interior walls creates a warm atmosphere, which changes according to the time and day and the positioning of the windows blinds.

Yiorgos Hadjichristou

REFURBISHMENT OF A LISTED HOUSE IN KAIMAKLI

> Nicosia, Cyprus | 2007 | Duration of project: 18 months | 2,809 sq ft | © C. Papantoniou, Y. Hadjichristou, Y. Kordakis <

The old village of Kaimakli is located in Nicosia, close to the so-called Green Line, one of the borders that divide Cyprus. This is the location of a project that consisted in remodeling a traditional house. The difference between the prosperity of the city of Nicosia and the careless atmosphere of the house's location creates a series of unexpected and surprising contrasts.

The conditions previously mentioned, combined with the objective of preserving the atmosphere of the house, generated the idea that the spaces should flow seamlessly. All the new spaces are organized around two courtyards and include mobile partitions. This mobility allows change and creates multiple combinations that can adapt to the time, desires and needs of the inhabitants. All the spaces have cross-ventilation and are south-facing, which protects them from direct sunlight in the summer months.

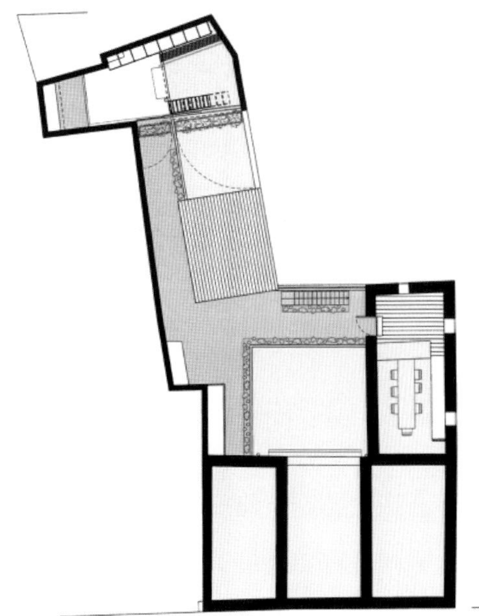

New second floor plan

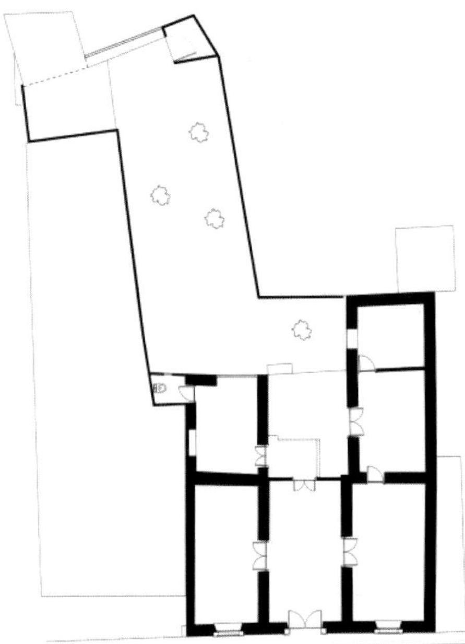

Existing floor plan

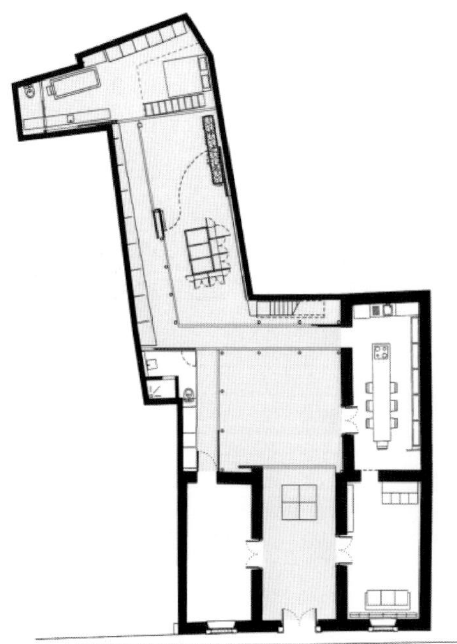

New ground floor plan

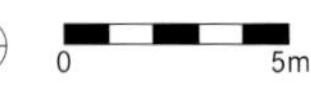

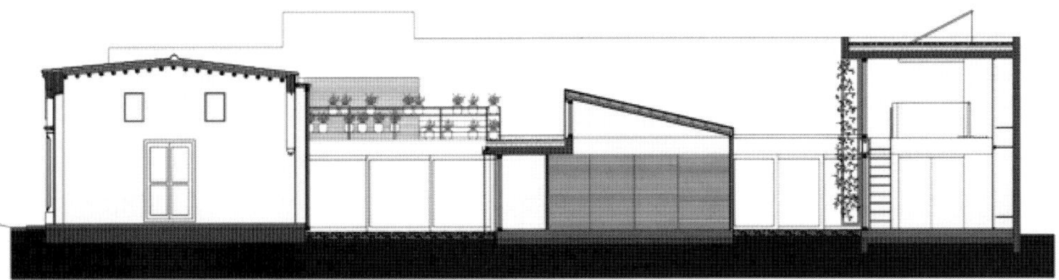

Section A–A'

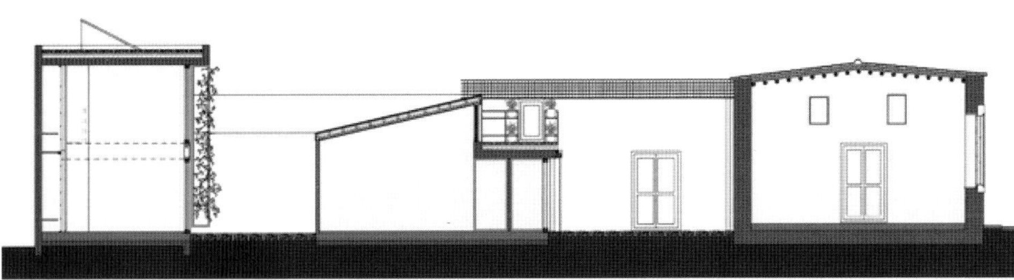

Section B–B'

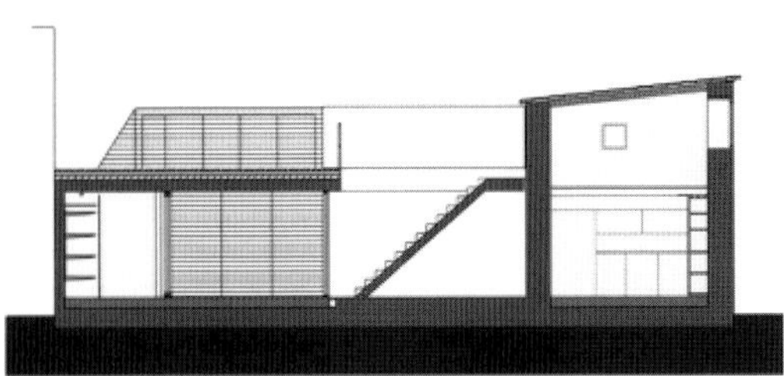

Section C–C'

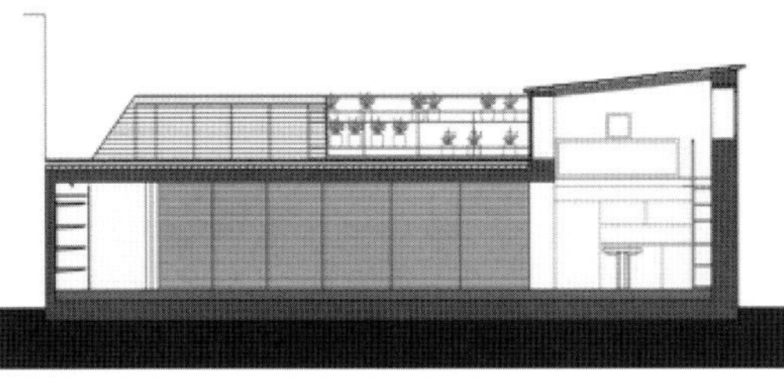

Section D–D'

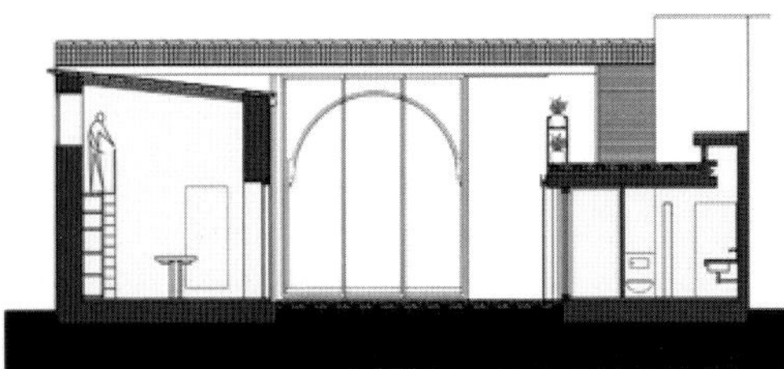

Section E–E'

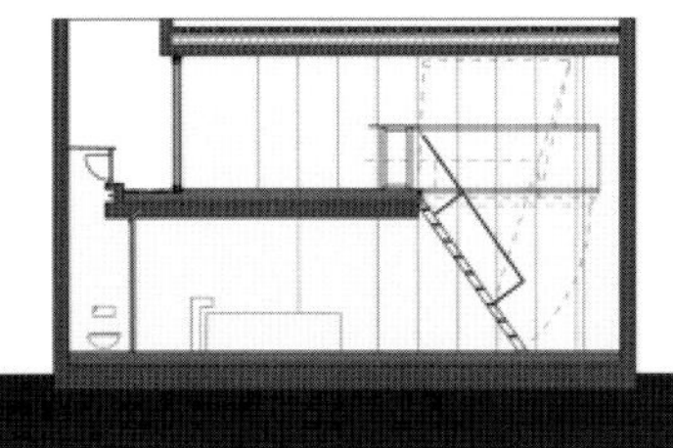

Section F–F'

In the renovation and extension of this Cyprian house, the architects followed the premise of respecting the dwelling's traditional structure and combining it with the modern character of the city.
The remodeling basically focused on distributing the rooms to generate spaces and cross-ventilation in order to make the most of the plentiful air in the area, since the property is south-facing, with its back against the strong summer sun.
These two characteristics are necessary to protect occupants from the Cyprus climate.

For most of the year, all the new mobile spaces are open to a courtyard in the center of the house. This area is called the sun room. The furniture also defines the space through rich and varied combinations that vary according to the time of the year and the needs of the occupants of the house. In this area, the mobile system looks like a Lego game suitable for all ages.

Mediterranean light plays an important role in the spatial design and solutions. The varying organization of the internal and external spaces is enriched with light that enters through the large windows and inner courtyards.

The terrace is the element unifying the new and old spaces in this traditional house. The same type of crushed stone used in the terrace is used in all exterior spaces as well as on the green roof. The mobile partitions are made from glass, polycarbonate and chipboard. The uncontrolled vegetation that previously invaded the house is revived in different parts of the property, used as another decorative feature.

> East Hampton, NY, USA | 2008 | Duration of project: 18 months | 3,560 sq ft | © Michael Lomont <

Stelle Architects

This 1920s house underwent a total makeover. The exterior of the house was renovated respecting the building's dimensions and the original architecture. The extensions, however, are the modern counterpoint to the original structure, while preserving the architectural composition and consistency prior to the remodeling works. In contrast, in the interior spaces, the architects wanted traditional and modern elements to happily coexist side by side. To achieve this, a minimalist and low maintenance design was chosen.

Due to the modest size of the house and in order to afford more privacy, a 860 sq ft guest house was added. The interior spaces were also given the insulation needed for the owners and their guests.

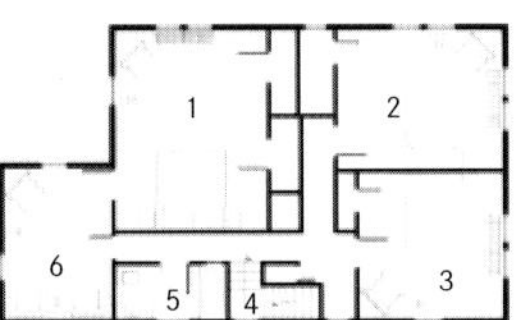

Existing second floor plan

1. Master bedroom
2. Bedroom 1
3. Bedroom 2
4. Hallway
5. Bathroom
6. Living room

New second floor plan

1. Master bedroom
2. Bedroom
3. Bathroom
4. Closet
5. Hallway

Existing ground floor plan

1. Closet	11. Wood deck
2. Guest bedroom 1	12. Dining room
3. Guest kitchen	13. Living room
4. Hallway	14. Entry porch
5. Guest bathroom	15. Entry hall
6. Guest bedroom 2	16. Kitchen
7. Storage 1	17. Pantry
8. Storage 2	18. Powder room
9. Brick patio	19. Vestibule
10. Guest parking	20. Driveway

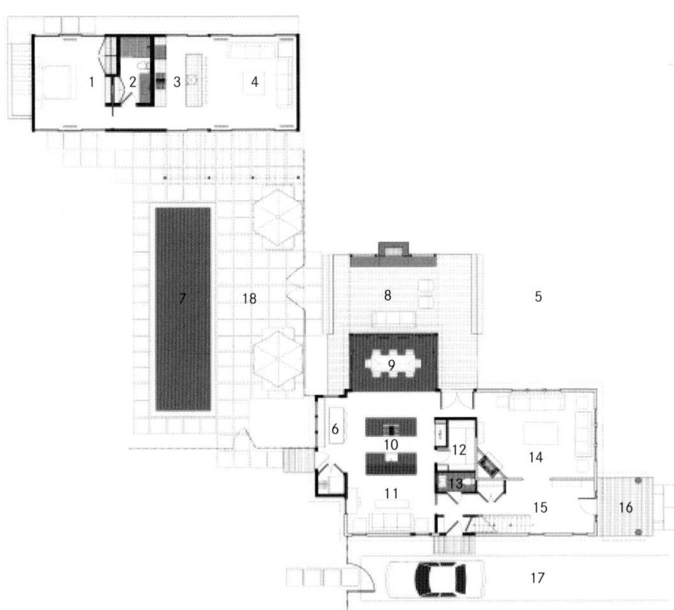

New ground floor plan

1. Guest bedroom	11. Family room
2. Guest bathroom	12. Pantry
3. Guest kitchen	13. Powder room
4. Guest living room	14. Living room
5. Lawn	15. Entry hall
6. Breakfast banquette	16. Entry porch
7. Pool	17. Driveway
8. Porch	18. Patio
9. Dining room	
10. Kitchen	

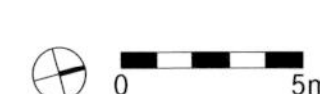

North elevation – main house

South elevation – main house

East elevation – main house

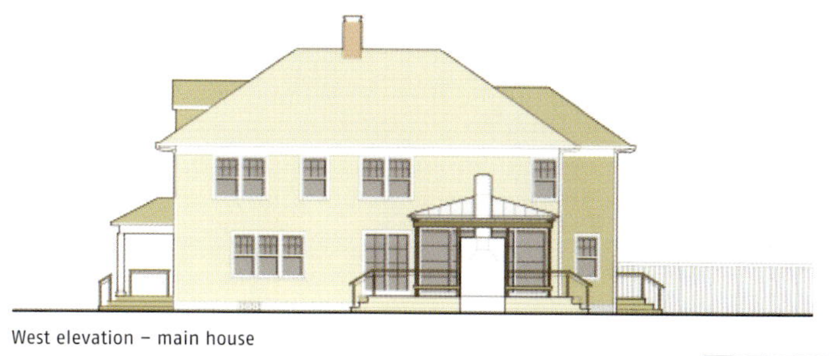

West elevation – main house

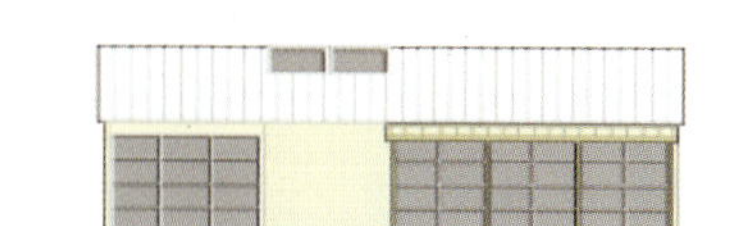

East elevation – guest pavilion

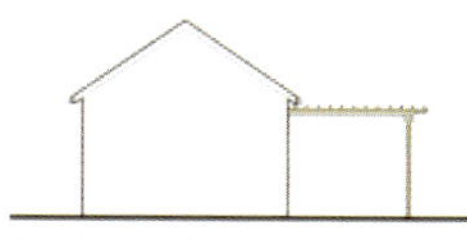

South elevation – guest pavilion

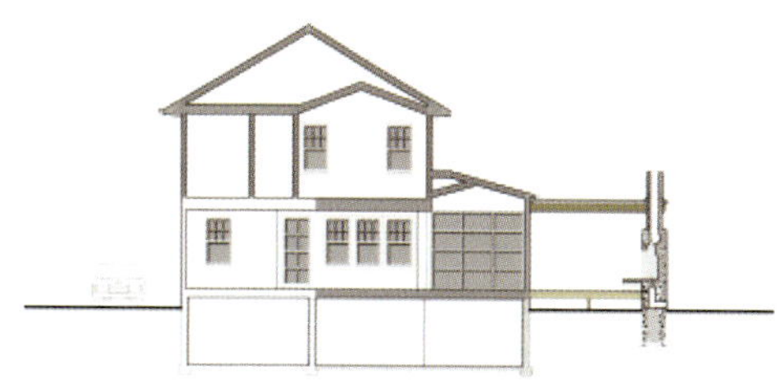

Section A–A' – main house

West elevation – guest pavilion

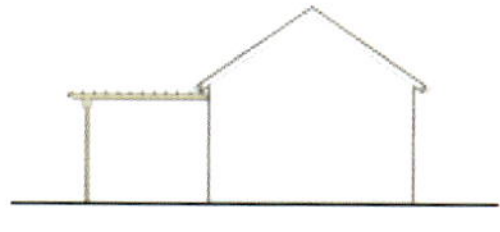

North elevation – guest pavilion

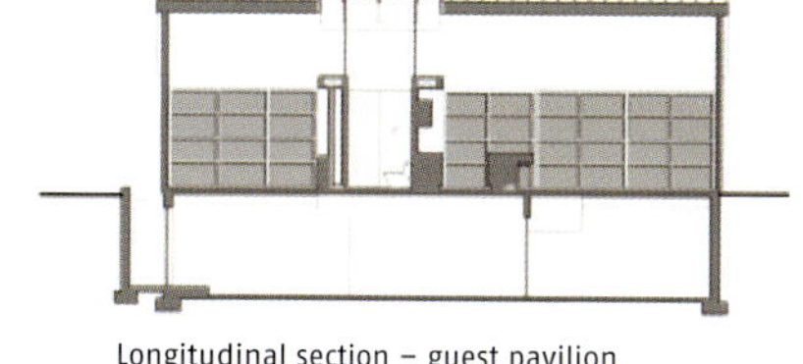

Longitudinal section – guest pavilion

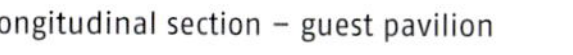

The result of the intervention on this residence was a two-story house where the most intimate spaces were located on the second floor and the common spaces, on the first floor. New connections were created between the two houses, and shared leisure zones were also created.

Besides building a new guest house, a new swimming pool, patio and barbeque area were constructed beneath a covered porch. The renovation was designed following sustainable criteria, such as using natural materials and installing geothermal air conditioning.

RESIDENCE 1414

> Austin, TX, USA | 2007 | Duration of project: 22 months | 5,658 sq ft | © Paul Finkel <

Miró Rivera Architects

The project's main objectives were to restore the exterior of this 1940s house, to transform its interior to give it more daylighting, and to improve its connection with the backyard. In the 1980s, the house was given a rather disagreeable extension, which the architects sought to remove.

The remodeling project used a simple palette of materials in order to maintain a balance between traditional elements of the original house and the modern reforms. Work was also carried out on the front door and in the front garden. The transformation sought to better connect the confined, gloomy interior spaces with the outside world. The backyard was integrated into the house, and the garage was remodeled to include a gym, bathroom and guest room in an upper level.

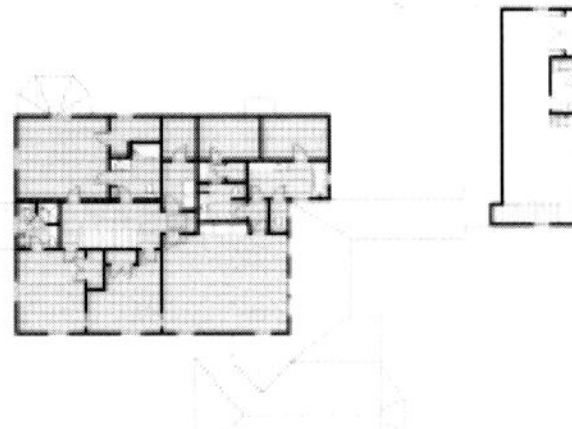

Existing second floor plan

New second floor plan

1. Hall
2. Bedroom
3. Bathroom
4. Laundry
5. Master closet
6. Master bathroom
7. Master bedroom
8. Office
9. Gym
10. Guest bathroom
11. Guest bedroom
12. Gym deck

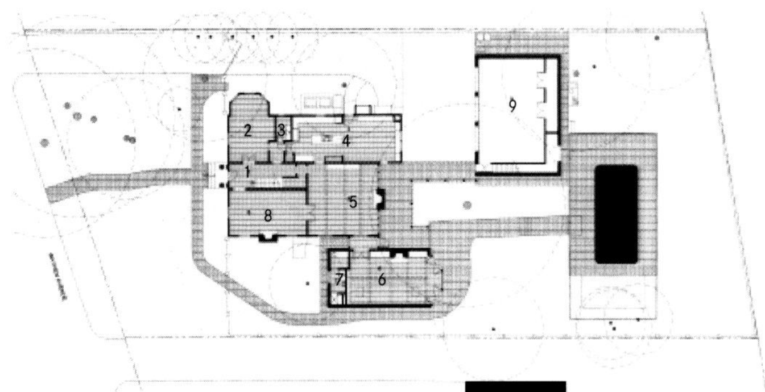

Existing ground floor plan

1. Entry
2. Dining room
3. Powder room
4. Kitchen
5. Family room
6. Den
7. Pool bath
8. Living room
9. Garage

New ground floor plan

1. Entry
2. Dining room
3. Powder room
4. Pantry
5. Kitchen
6. Family room
7. Den
8. Pool bath
9. Side patio
10. Living room
11. Garage
12. Backyard patio

0   5   10m

The main aim of the architects was to create interaction between the exterior and interior spaces. Before the renovation, the house was inward-facing, with large expanses of walls and small windows that did not communicate with the various patios surrounding the house.

Outside, the range of colors and materials were selected to maintain a balance between the old and new. Traditional features, dating from the 1940s, have been recovered and perfectly combined with new ones, as can be seen in the main entrance.

The dwelling's interiors were gloomy and confined so the interior transformation focused on increasing natural daylighting. The white walls and ceilings emphasize luminosity and play with the wood color of the furniture and floor.

HOUSE DE BELDER-ROBIJNS

> Heverlee, Belgium | 2005 | Duration of project: 2 years | 9,171 sq ft | © Nullens André <

BOB361 Architects

This project is located in Heverlee, near Brussels, in one of the town's main streets. Over time, Flemish-style annexes were added to these two-story houses.

In the remodeling, the first act was to demolish the extensions. The aim was to recover the original state of the houses, just as they were designed in the beginning. Through removing these structures, the architectural complex was once more filled with light and air.

To blend modernity and tradition, large glazed modules were created. On the front façade, these glass blocks act as a screen, while defining a terrace that connects with the kitchen. This type of module was also installed to the rear of the house, connecting to the patio. This original structure affords greater privacy from the neighboring houses and lets more natural light into the interior spaces.

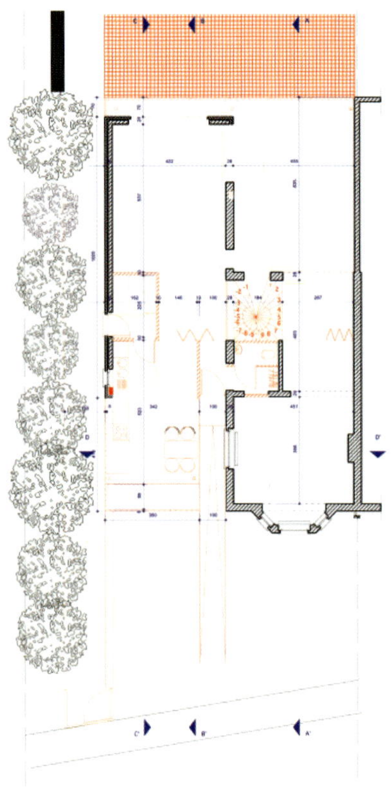

Existing ground floor plan

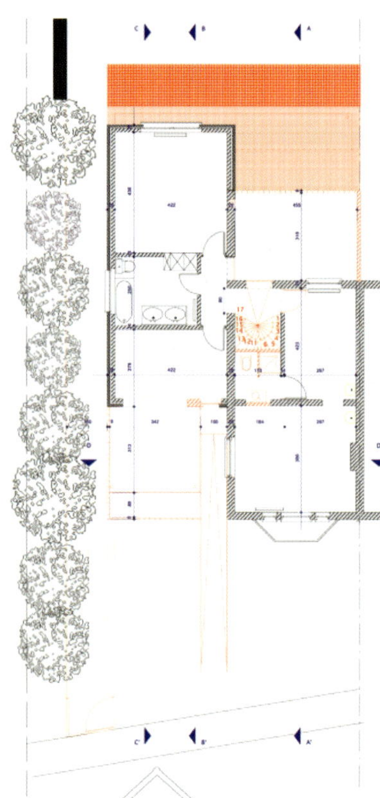

Existing second floor plan

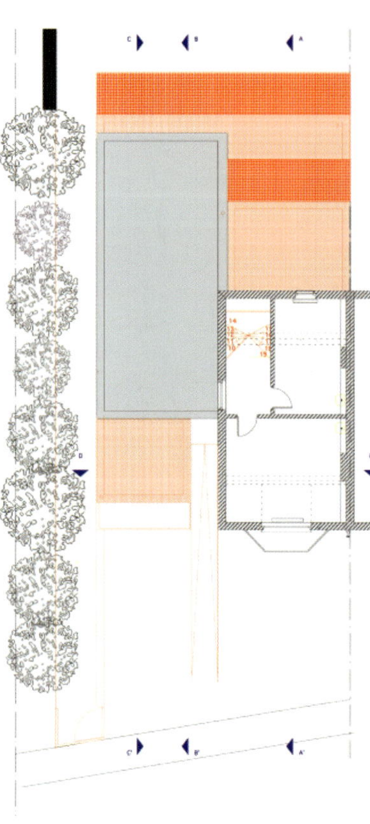

Existing third floor plan

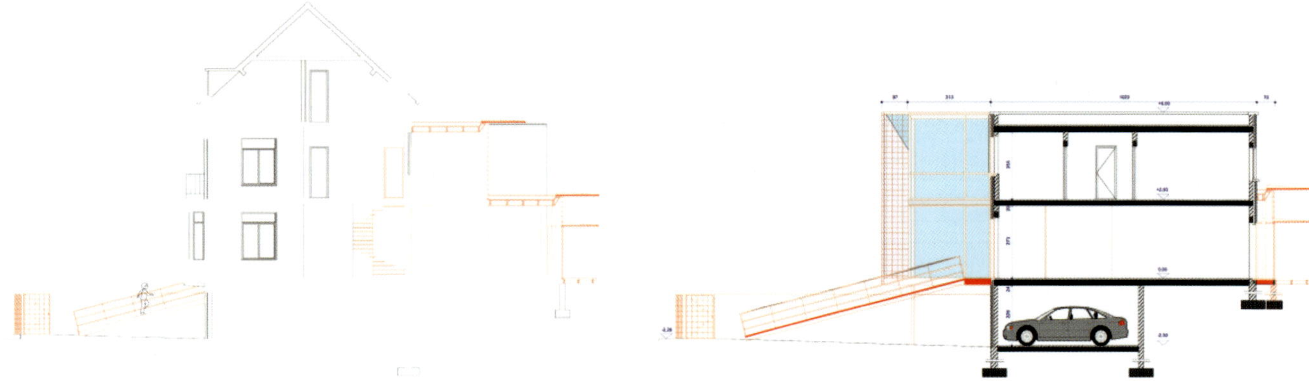

Existing section A–A'

Existing section B–B'

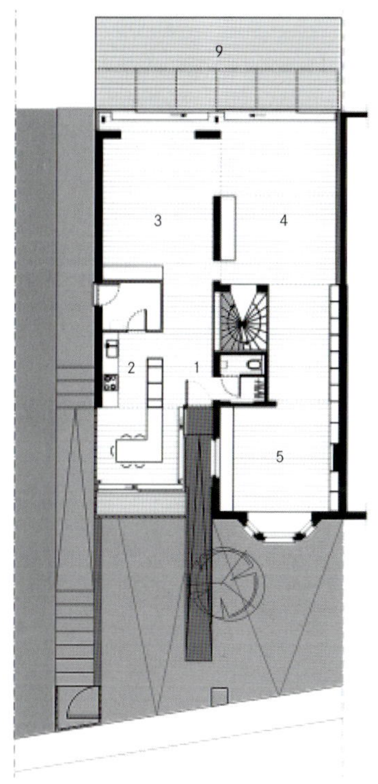

New ground floor plan

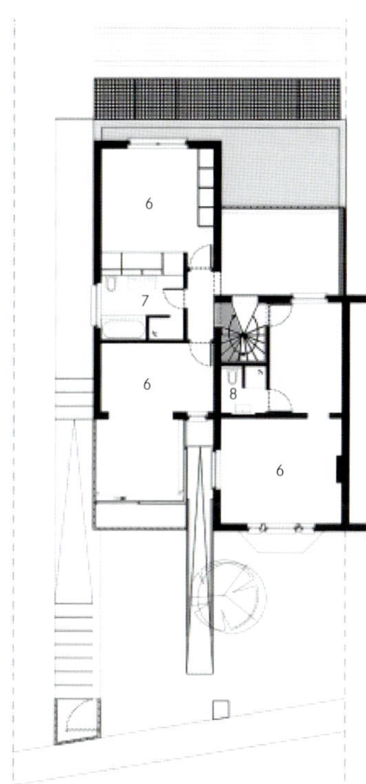

New second floor plan

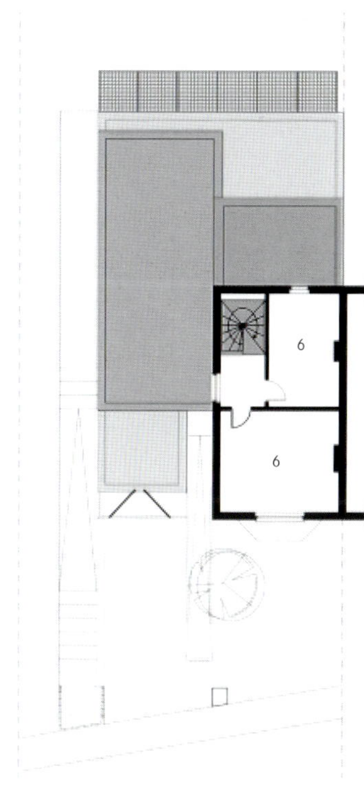

New third floor plan

1. Vestibule
2. Kitchen
3. Dining room
4. Living room
5. Studio
6. Bedroom
7. Bathroom
8. Shower room
9. Terrace

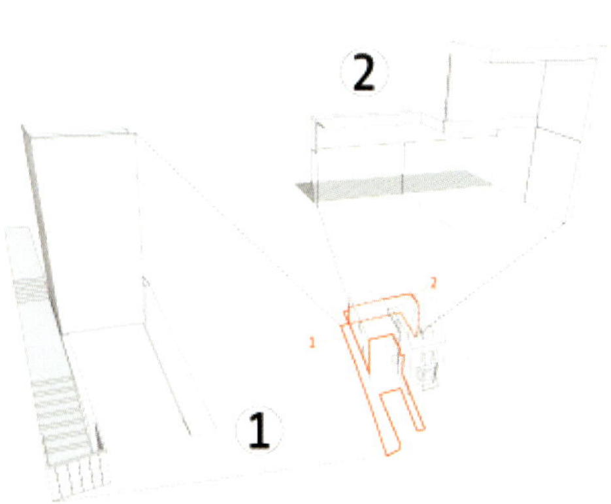

Exploded perspective of intervention

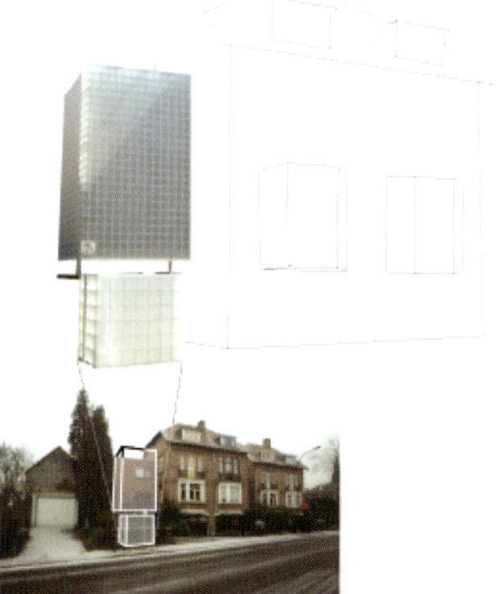

Intervention photomontage

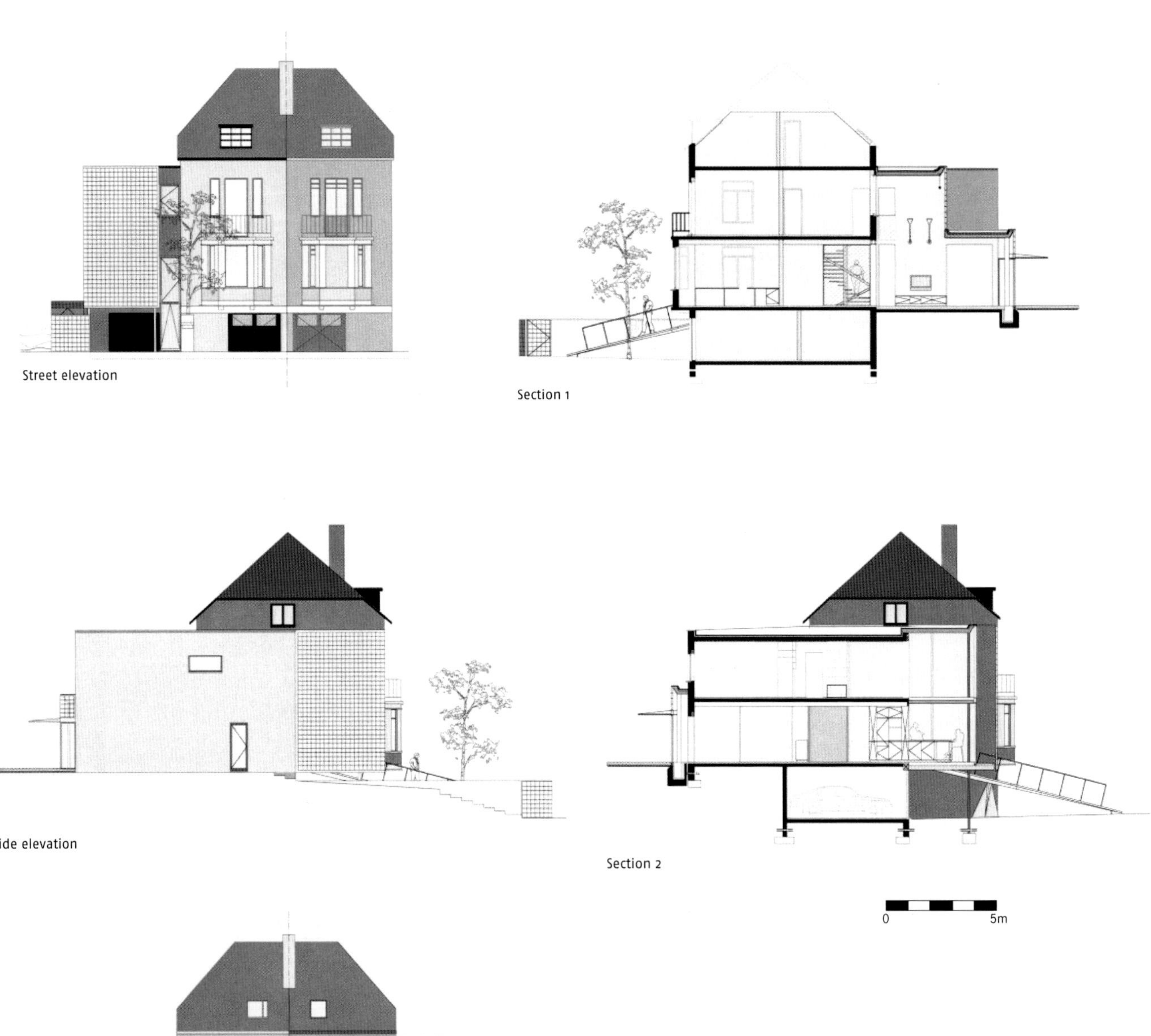

Street elevation

Section 1

Side elevation

Section 2

0   5m

Back elevation

An open, compact staircase was created to access the glazed addition. The new structures make the interior a unique and seamless space. They are designed and built using large prefabricated components. These spectacular and original walls are formed from white concrete and translucent glass.

By creating these new structures, the architects played with the idea of seeing and not seeing, and being inside and outside. In the main façade there is a two-story rectangular form, whose interior is sensed but can not be seen. Another module, completely open and glazed, with the same characteristics but this time horizontal was built at the rear of the property.

The precedent forms and new additions are perfectly integrated. The new building is fully glazed to set up a dialogue with the brick building. This second "skin" made from white concrete and translucent glass gives the clients the privacy they required. The flooring consists of wooden slats that have been inset with lights, which create a dramatic effect when there is little sunlight.

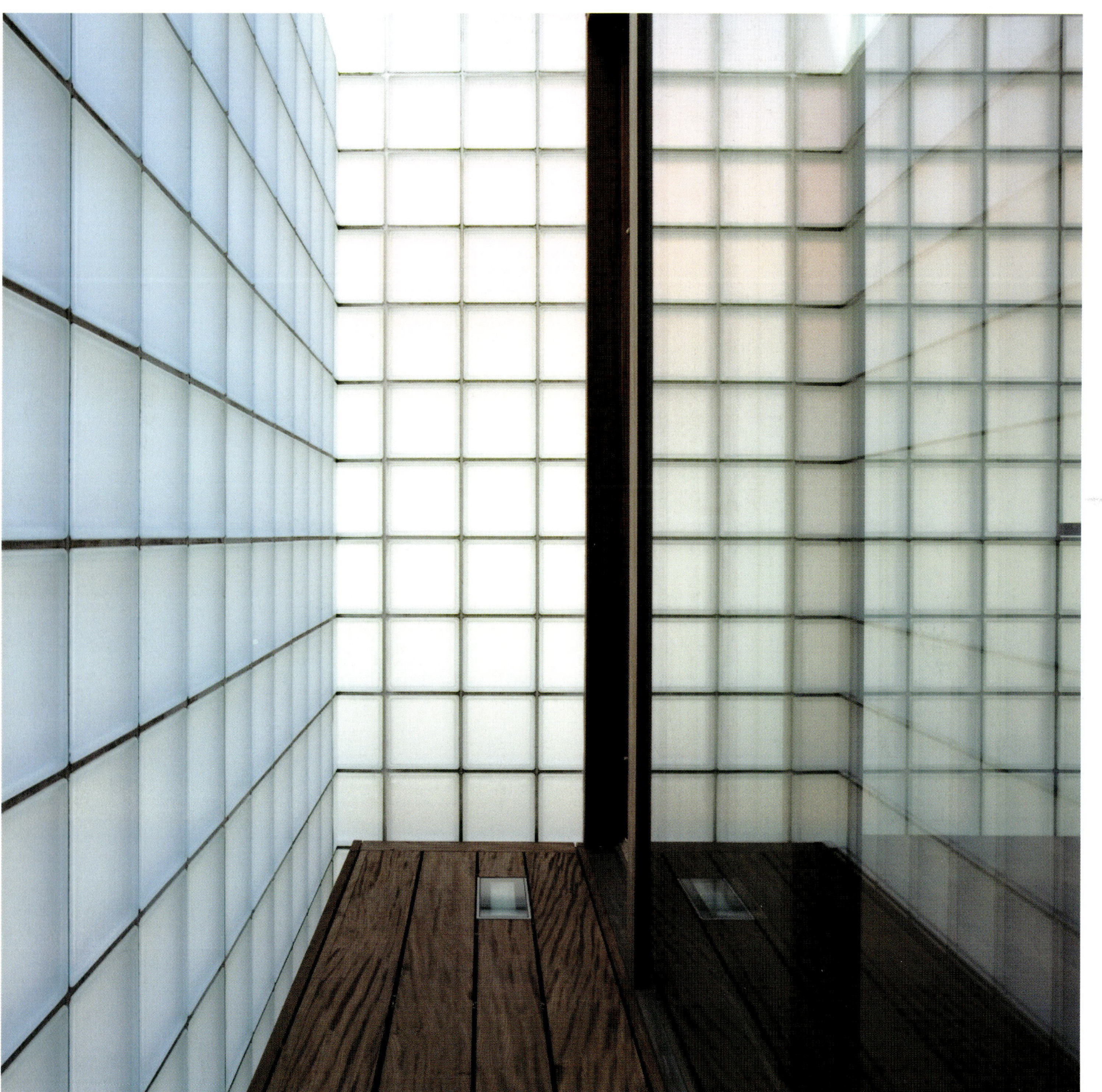

Edward Szewczyk & Associates Architects

> Woollahra, NSW, Australia | 2006 | Duration of project: 18 months | 3,229 sq ft | © Michael Saggus, Ute Wegman <

In the 19th century, behind his small house in the historic center of the suburb of Woollahra in Sydney, the well-known bacteriologist John McGarvie Smith set up a laboratory to study the anthrax vaccine. The two buildings, both the cabin and the laboratory, have been conserved over the years, and the owner decided that these should be incorporated into the new dwelling.

The client's brief was to construct a house that would provide a high quality of living through a functional and contemporary design. This was achieved through establishing an interaction between the old buildings and the new additions: the value of the textures and the details of the old buildings were highlighted with the expressed contained in the new forms.

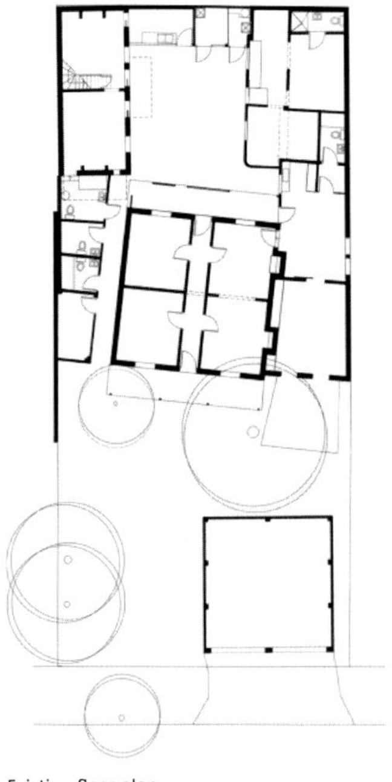

Existing floor plan

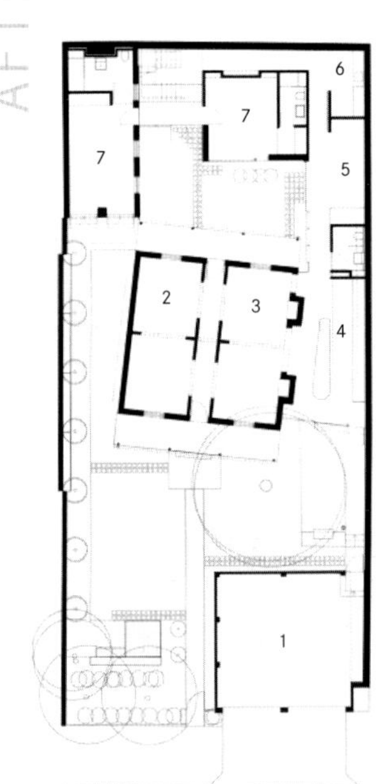

New ground floor plan

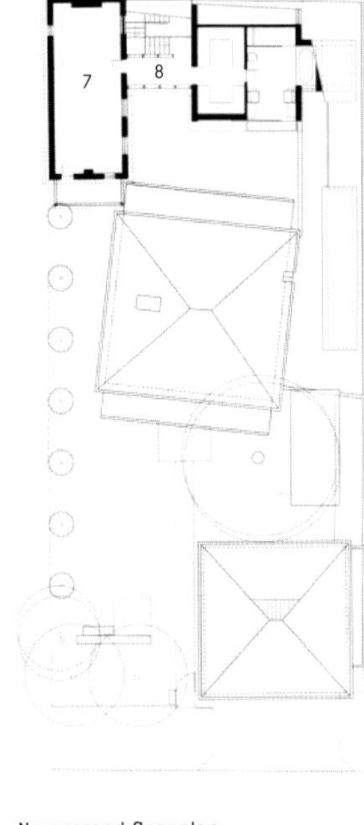

New second floor plan

1. Garage
2. Living room
3. Dining room
4. Kitchen
5. Study
6. Laundry room
7. Bedroom 1
8. Bridge

Cross section

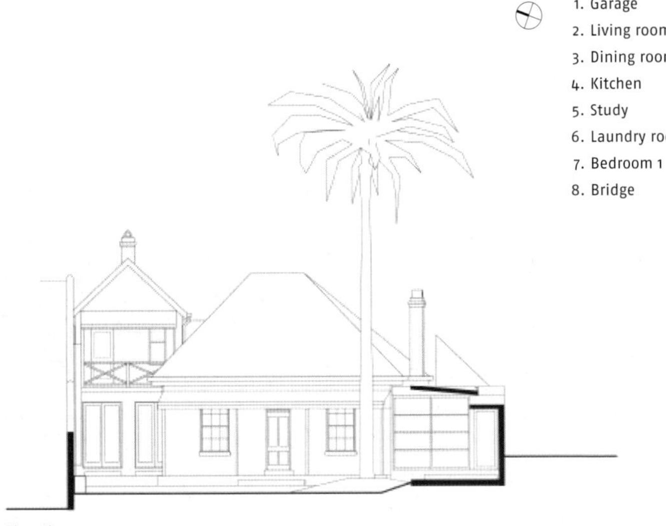

Elevation

The new additions are reorganized around an atrium located in the edge formed by a line of trees. In this way, the old constructions are interconnected. This provides a long internal corridor that offers numerous viewpoints of the city. When the inhabitants move around the intercommunicated spaces, they will be able to appreciate the variety of forms of the building, depending on their viewpoint.

The remodeling program was based on the premise of conserving the old buildings and integrating them into the new additions to create a single complex. Through this design, the client wanted to reflect modern living. To achieve this effect, classic elements have been reused and combined with contemporary furniture. Minimalism is evident in some of the interior spaces, such as the kitchen.

The existing ornamental elements are emphasized and new contemporary designs are created in a harmonious manner. Artificial lighting has been located in an ad hoc manner, such as in the bathrooms and in the kitchen, since most of the rooms have large windows that bathe the interior spaces with daylighting for most of the day.

OPEN BOX 2

\> San Francisco, CA, USA | 2007 | Duration of project: 18 months | 2,200 sq ft | © Paul Dyer <

Feldman Architecture

The project follows the same design concept conceived in the previous project of the architects: the Open Box House. This time, the studio focused on transforming a two-story family home with basement and garden. Despite radical modifications to the interior spaces and main façades, the aim was to maintain the original footprint of the house.

This renovation helped make the dwelling a more habitable space and improved its relationship with its surroundings, taking into account environmental factors such as radiant heating and rooftop solar panels. Another important component in the remodeling consisted in fitting windows, which maximize daylighting without compromising the building's privacy with respect to the rest of the houses, and large French doors that access the garden from the bedrooms. The end result is an open space with sophisticated finishes that give the dwelling a modern character.

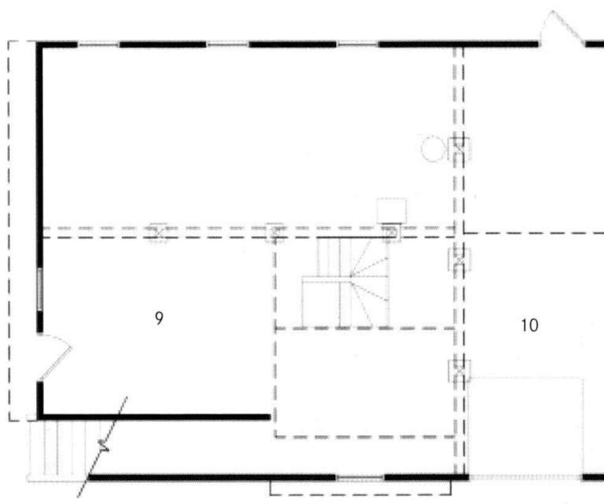

Existing second floor plan

1. Hall
2. Living room
3. Dining room
4. Kitchen
5. Bathroom
6. Bedroom 1
7. Bedroom 2
8. Bedroom 3

Existing ground floor plan

9. Basement
10. Garage

The architects carried out several interventions focused on setting up a dialogue between the interior and exterior spaces. Prior to the remodeling, the structure had few openings in the façade. The project carefully positioned the windows to provide sufficient natural lighting while also giving the privacy required by the client.

AFTER

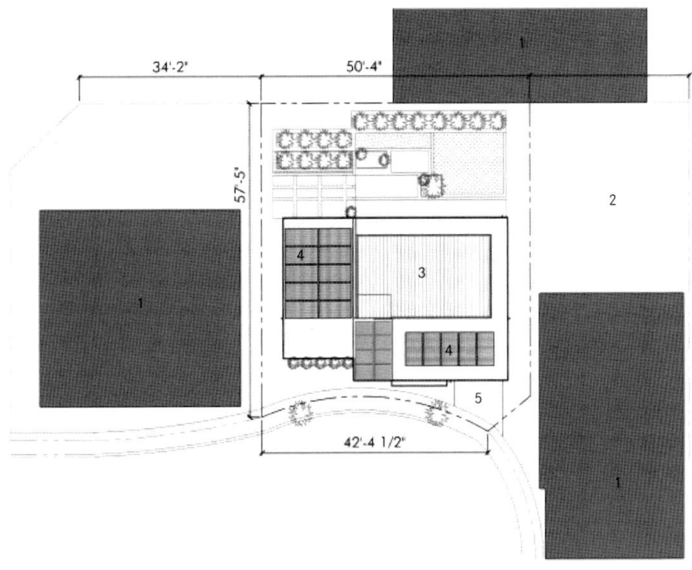

Site plan

1. Adjacent building
2. Garden
3. Deck
4. Solar panels
5. Roof access skylight

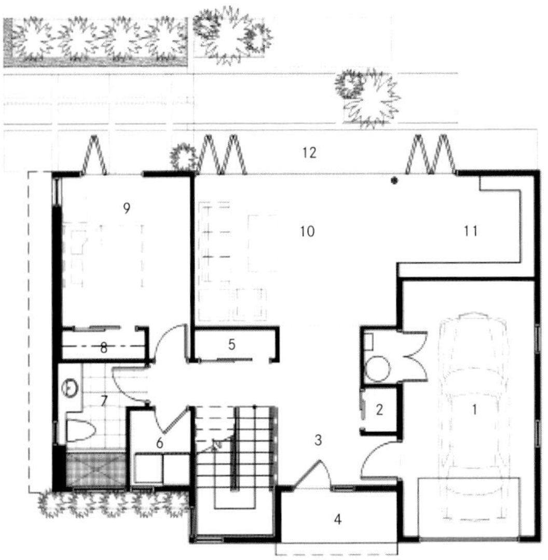

New ground floor plan

1. Garage
2. Coats
3. Entry
4. Entry porch
5. Storage closet
6. Laundry room
7. Bathroom 2
8. Closet 1
9. Bedroom 2
10. Family room
11. Study
12. Courtyard

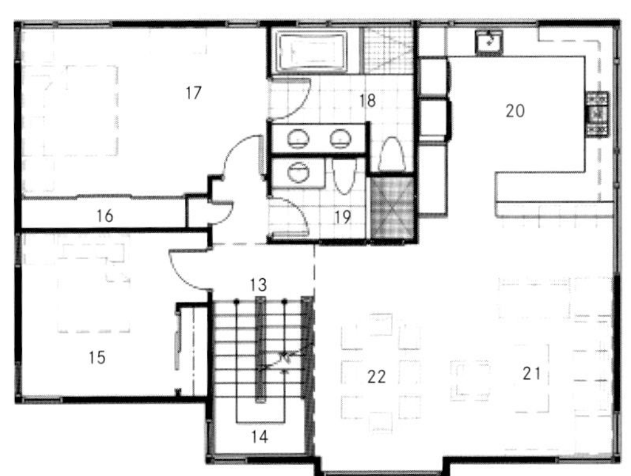

New second floor plan

13. Stairs
14. Closet 2
15. Bedroom 1
16. Closet 3
17. Master bedroom
18. Master bathroom
19. Bathroom 1
20. Kitchen
21. Living room
22. Dining room

Although the renovation brief was to segment this 1940s house, the real challenge proved to be amplifying the space without altering its previous state or removing a good part of the exterior walls. The Open Box 2 was modernized with elegant finishes such as the Italian oak furniture, the fine hardwood flooring, the shiny tiles that add texture to the kitchen walls, and finally, the xilo tiles in the bathroom. Another important feature is the breakfast bar.

Large French doors give access lead out to the garden from the bedrooms on the lower floor. This is an example of the architects' aim to bring the outside inside, and vice versa. An automatic door, like a trapdoor, has been installed on the roof. This functions as a skylight and leads out onto the roof with views over San Francisco.

> Surf Beach, Australia | 2008 | Duration of project: 1 year | 4,381 sq ft | © Tony Trobe <

Tony Trobe/TT Architecture

This project could be subtitled "vista", as the view is both the lot's main characteristic and its prevailing design feature.

This intervention, which started out as a renovation, was the result of a various proposals put forward by other designers. The clients, who showed an appreciation for the taste of the design, participated enthusiastically, and were actively involved in the design's gradual evolution. They were also generously flexible in the creative process, allowing a powerful architecture to be manifested in the renovation.

The designers used vertically aligned masses and left them slightly curved; these two actions, combined with a minimalist approach led to the creation of a very modern and spectacular house.

The dominant shape in the roofs and the organization of the floor spaces were sufficiently powerful to prevent a single color scheme from interfering with the building's architectural expression.

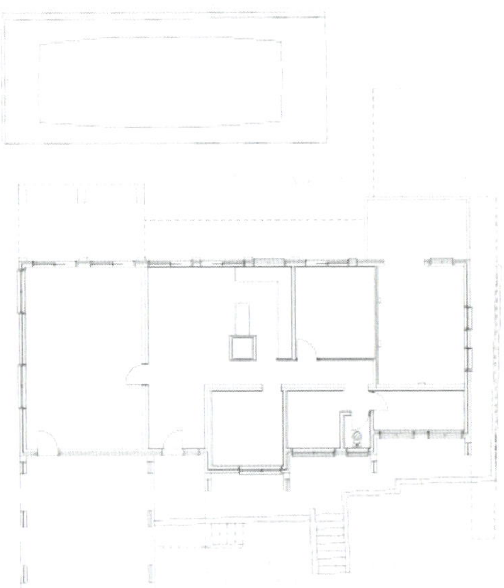

Existing upper floor plan

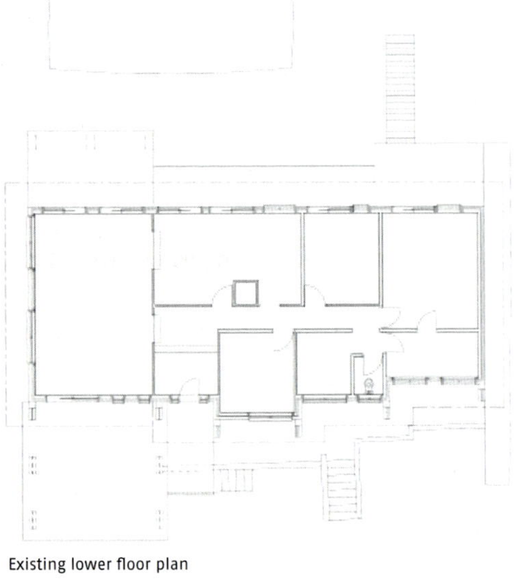

Existing lower floor plan

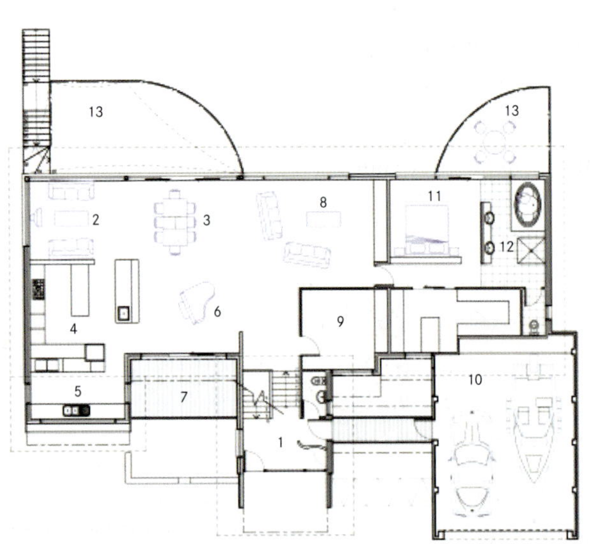

New upper floor plan

New lower floor plan

1. Entry
2. Breakfast nook
3. Dining room
4. Kitchen
5. Pantry
6. Entertainment area
7. Deck
8. Living room
9. Storage
10. Garage
11. Bedroom
12. Bathroom
13. Patio
14. Laundry room
15. Bar
16. Pool

Although big windows were designed to capture natural light during the day, the architects also designed a system of artificial lighting that provided focal points of lighting. Spotlights were strategically located to give the house a monumental aesthetic.

One of the key features of this renovation was the creation of spectacular roofs that cantilever over the main body of the house. Both the exterior and interior spaces have a minimalist design scheme.

The unique, sloping forms of the land enabled the creation of spaces that interact with their immediate surroundings. For this reason, the communal areas were created on the upper level. The many terraces, the pool, and the living and dining rooms were all designed to make the most of the lot's spectacular views.

Architects EAT

> Cremorne, VIC, Australia | 2004 | Duration of project: 15 months | 1,830 sq ft | © Architects EAT <

The house is an addition located in a relatively small space, in an area where houses are built occupying the total area of the site. The urban environment is very dense and is characterized by houses that overlook the lot. The remodeling project therefore focused on providing the building with natural light and cutting it off from the rest of the neighboring houses to give the clients the privacy they required. One technique used was creating pillars that housed French doors giving access to the garden.

The main work was carried out on the roof, where three openings were made in the form of linear skylights across the width of the surface. This original solution allows maximum sunlight to enter and creates interesting lighting effects. There is also a skylight in the bathrooms, above the shower, which allows occupants to be bathed in sunlight while enjoying a soak.

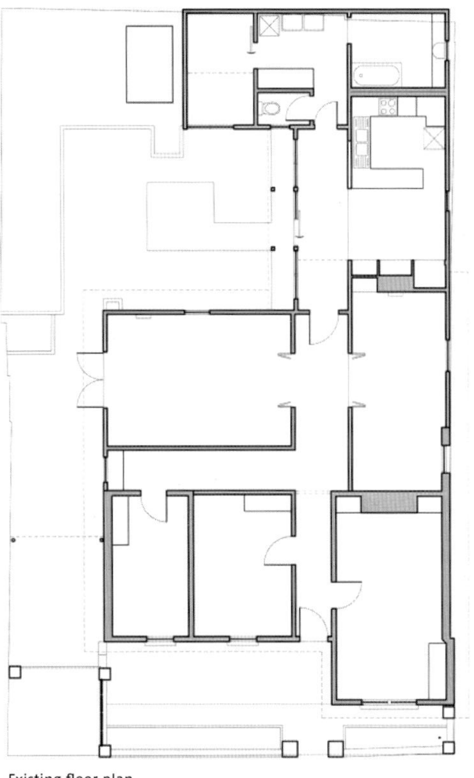

Existing floor plan

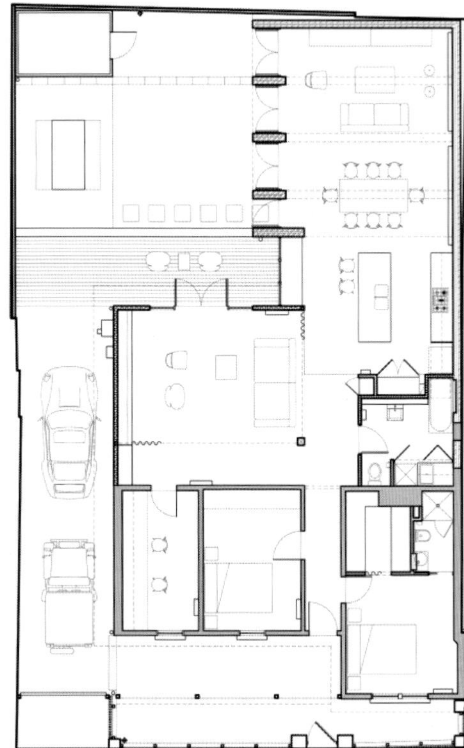

New floor plan

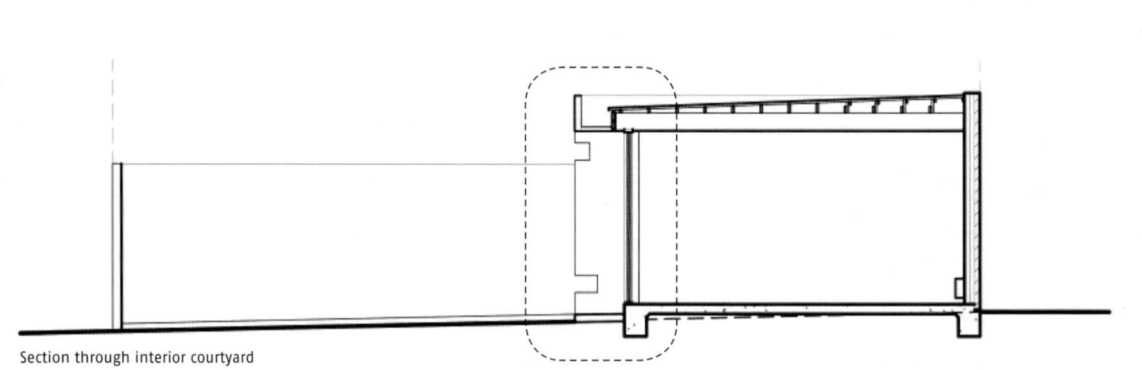

Section through interior courtyard

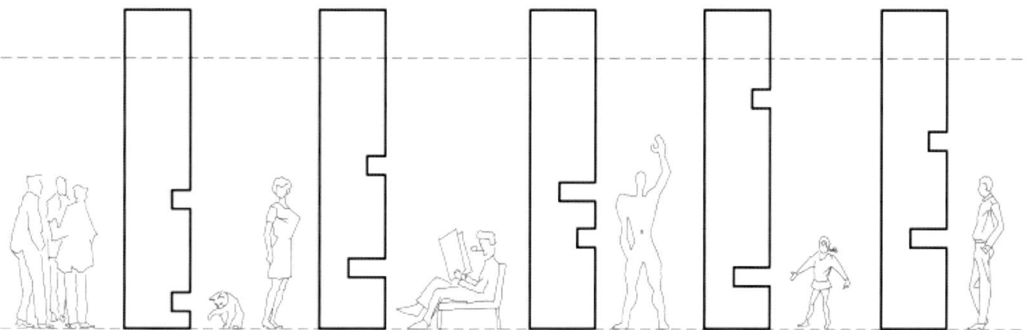

Conceptual design of fin sculptures

Elevation at fin sculptures

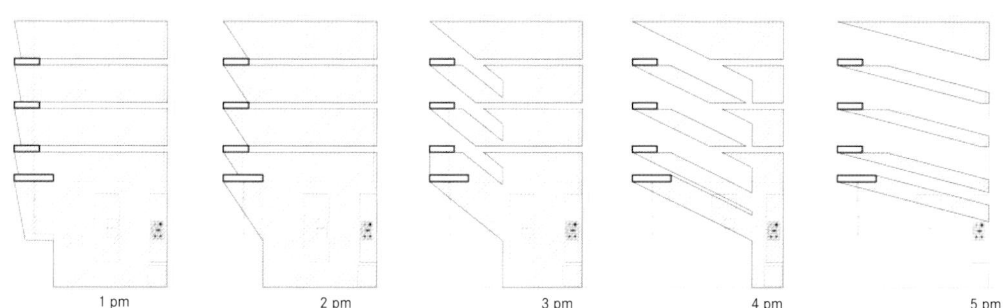

Shadows study at fin sculptures

The option to create large glazed expanses overlooking the garden was ruled out due to the proximity of neighboring houses. Five vertical pillar-like elements were created instead, partially blocking the views of neighbors. These elements divide the façade into four openings where doors are installed giving access to the garden. These double as structural elements supporting the roof.

The five pillars framing the French doors were one of the most important design creations. Original cuts were made in these at different heights and with varying depths. This feature gives them an almost sculptural plasticity. The original idea was to create practical shelving for wine glasses, ashtrays and plants, but they soon became a feature that led to an exploration of materials and shapes.

The kitchen is designed to be the nerve center of the house, since it connects the living–dining space with the study–bedroom. It is a place where the occupants spend most of their time, either cooking or entertaining guests. This central position gives panoramic views of the house and garden through the slatted windows.

The colors and finishes were
selected to serve as intermediately
elements: black predominates in the
study–bedroom, and white plays
a central role in the living–dining
space in the heart of the house.
Air circulation is obtained through
louvers in the windows that also
regulate daylighting.

JAM

Apollo Architects & Associates

> Tokyo, Japan | 2009 | Duration of project: 1 year | 2,669 sq ft | © Masao Nishikawa <

This 40-year-old residence had undergone several very different remodeling projects. The new architects came up with a minimalist design that combines floors and ceilings in dark tones with pale walls and furniture.

The house consists of a ground floor, where some of the common spaces are located, such as the garage and entranceway; a first floor with kitchen, living room and private areas, and finally a second floor with a bathroom, bedroom and a furniture-free room.

One of the most successful interventions was the main façade. The size and layout of the windows were used, but these were projected towards the exterior. The frames were painted black, while the rest of the façade is white. This color scheme follows the criteria that the architects used in each of the rooms of the house. Projecting the frame of the windows creates a varying effect of light and shade in the interior design scheme.

BEFORE

AFTER

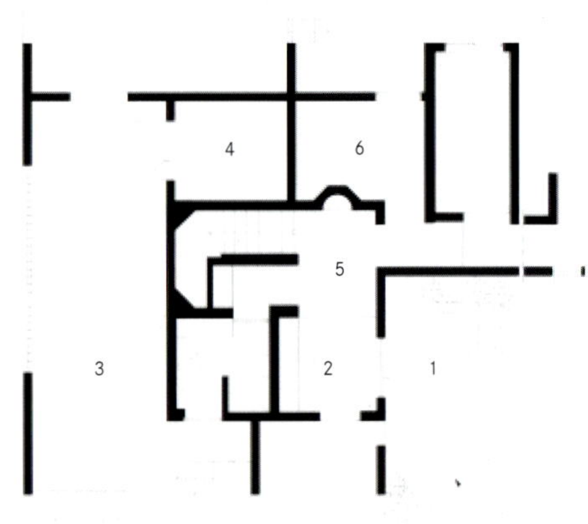

1. Porch
2. Entry
3. Garage
4. Machine room
5. Study room 1
6. Hall 1

New ground floor plan

Existing ground floor plan

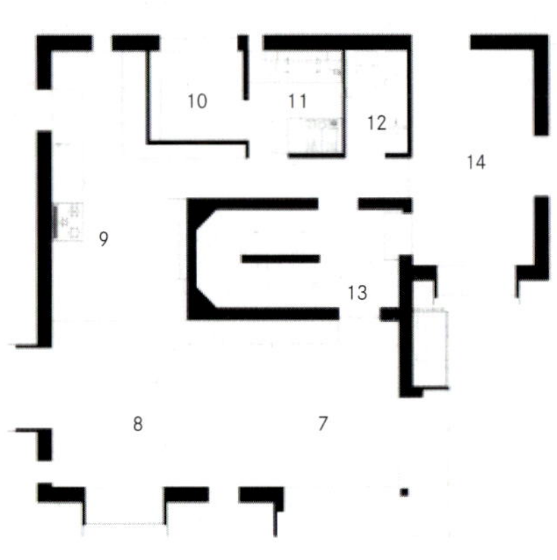

7. Living room 1
8. Dining room
9. Kitchen
10. Bathroom 1
11. Powder room
12. Bathroom 2
13. Hall 2
14. Study room 2

New second floor plan

Existing second floor plan

Existing third floor plan

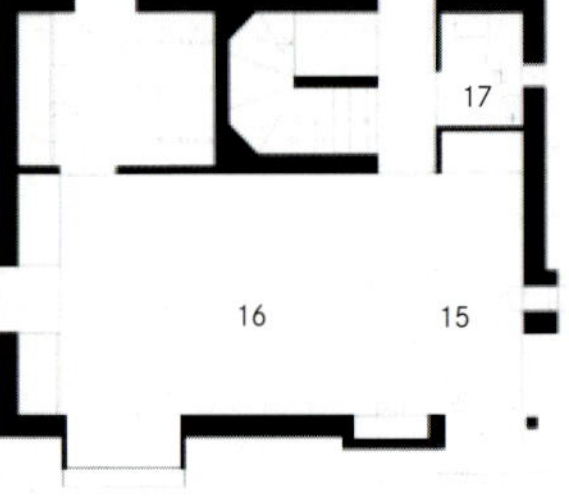

15. Living room 2
16. Bedroom
17. Bathroom 3

New third floor plan

One of the most original features of this project is the design of the windows. Although the architects maintained their original position, they designed two types of windows: the first projecting out towards the exterior through the frame and others that sit flat against the façade. From inside, the windows projecting outside provide greater depth. The windows located in one of the corners of the house, both in the first and second floor, allow better daylighting.

The contrast generated by this color scheme, between the white of the walls and the dark flooring, was one of the objectives of the architects. The sparse furniture combines dark tones on the first floor and pale tones on the second.

The lighting is indirect and inset in the furniture. This type of lighting gives the house a more visual dimension. The dark flooring emphasizes the light and shade effect: as the light hits the floor, rectangular shapes are created.

Ian Ayers & Francesc Zamora

HOMAGE TO A FILM MAKER

> Barcelona, Spain | 2009 | Duration of project: 3 months | 754 sq ft | © Francesc Zamora <

This renovation on a slim budget was carried out by a young writer in an apartment he had recently purchased in a new block in the 22@ Barcelona district. Simple, original finishes were typical for this type of building.

As the kitchen had been relegated to a narrow, gloomy nook, firstly, the wall between the kitchen and the dining-living room was knocked through to create a much more ample space. This also freed up space to place a bar against the kitchen cupboards to create a more practical design.

Secondly, a bright range of colors were brought into play. The two colors in the kitchen and dining-living room intermingle, while the hall, corridor and bedrooms were painted with vertical stripes and vibrant colors. All the colors used in the flat were painted on a narrow section of wall next to the front door; this is the first feature that catches the eye upon entering the apartment.

Existing floor plan

New floor plan

1. Entrance hall
2. Living room
3. Balcony
4. Kitchen
5. Laundry room
6. Hallway
7. Bedroom
8. Bathroom
9. Master bedroom

1. Entrance hall
2. Living room
3. Balcony
4. Kitchen
5. Laundry room
6. Hallway
7. Bedroom
8. Bathroom
9. Master bedroom
10. Bookshelves
11. Dining table
12. Kitchen island
13. Storage unit
14. Closet

Technicolor-inspired color scheme

Technicolor-inspired surfaces and stripes of color are a simple yet effective theme in this design scheme. The floor plan has only undergone small changes: only the wall between the kitchen and the dining-living room has been removed to provide more daylighting through the window located in the latter space. A new lighting system sets off the colors.

The kitchen and dining–living room are connected through two different colors. The grayish purple satin over the suspended ceiling and the kitchen walls are complemented by the greenish yellow in the living-dining area. Both colors intermingle: the dining room color invades the kitchen area and vice versa.

The corridor connecting all the rooms in the flat is an enclosed, windowless space. It was lit and revitalized with stripes and a bright red color, echoing the color scheme used in the kitchen and dining-living room. The white baseboard and the door frames create a striking contrast.

OS
OS
PRIETO A.S.C., A.M.C.
ALBERTO IGLESIAS
ontaje: JOSÉ SALCEDO
la por: ESTHER GARCÍA
AGUSTÍN ALMODÓVAR
ta y dirigida por:
MODÓVAR
cameo
Avid

X-LOFT

> Roma, Italy | 2007 | Duration of project: 8 months | 1,184 sq ft | © Filippo Vinardi <

Before this old Rome apartment was renovated, its layout consisted of a series of rooms running off from a gloomy corridor. The renovation transformed the corridor into an elongated, illuminated space, which offers a variety of interesting perspectives through a structure of beams and columns.

Colors played a central role in the remodeling; red was used to connect the different areas of the house and give a strong character to its common areas.

This loft is designed as a place with seamlessly flowing spaces, guided by the red structure of beams and columns. The kitchen was incorporated into the structure to achieve a powerful fusion between architecture and furniture design.

Special care was taken when choosing the materials to create a contrast between the warm colors and textures, and the industrial style.

SPECIAL 25th ANNIVERSARY PRESENTATION
The Godfather
CASABLANCA

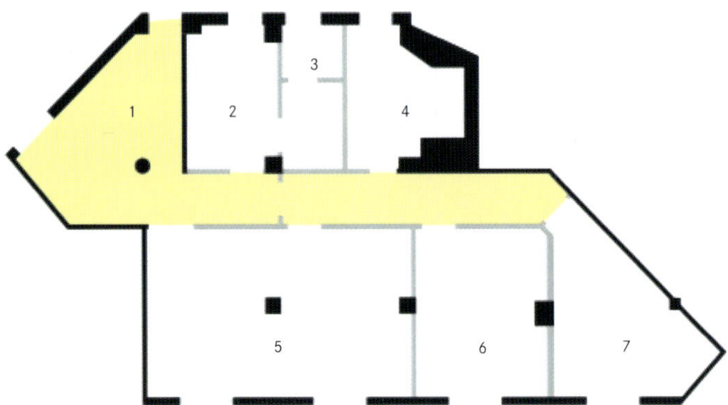

Existing floor plan

1. Entry
2. Kitchen
3. Bathroom 1
4. Bathroom 2
5. Living room
6. Bedroom 1
7. Bedroom 2

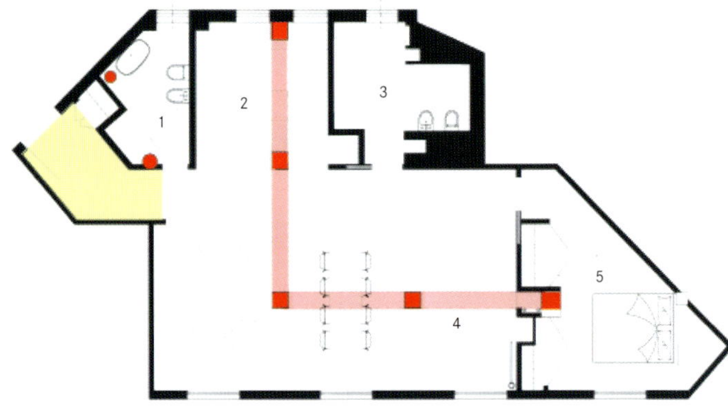

New floor plan

1. Bathroom 1
2. Kitchen
3. Bathroom 2
4. Living room
5. Bedroom

Model 1

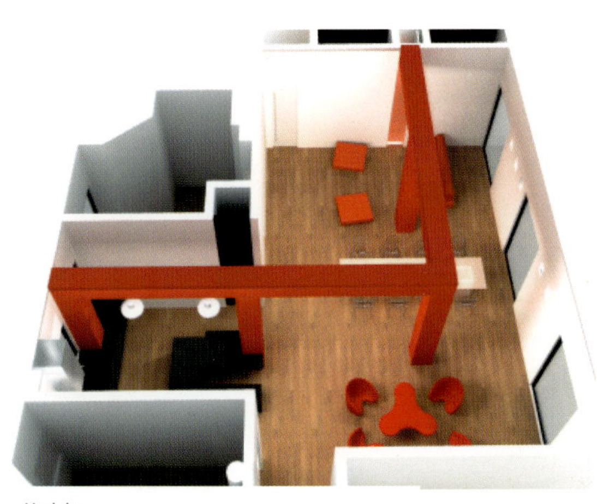

Model 2

The kitchen was designed to open onto the living room, in line with the house's architectural design concept. The furniture was designed by the same architect and reinforces the loft's minimalist style Black oak was used to lessen the boundaries of the kitchen so that it is perceived as part of a continuum.

(Next pages). The master bedroom has been separated from common areas by a double-skin partition that houses different bathrooms. The guest bathroom, formerly the house's entranceway, has been adapted to the building's irregular shape and structure in order to create a modern, industrial style.

> Zaragoza, Spain | 2007 | Duration of project: 8 months | 1,066 sq ft | © Eugeni Pons <

Magén Arquitectos

The owners wanted to completely reform the interior of this apartment, located on the third floor of a historical building. Prior to the remodeling works, the apartment was excessively fragmented in a linear distribution of a corridor and rooms.

The project consisted in emptying the interior space and adding furniture units to divide the void into areas for different uses. This solution dispensed with the need for interior partitions, opting instead for an open-plan, flexible and luminous space. One of the architects' main objectives was to create a visual continuum from one end of the house to the other.

The interior space is divided into three consecutive zones: the entrance area —which includes the reception and open-plan kitchen—, a rest area, and finally the living-dining space. The latter is the focal space in the layout.

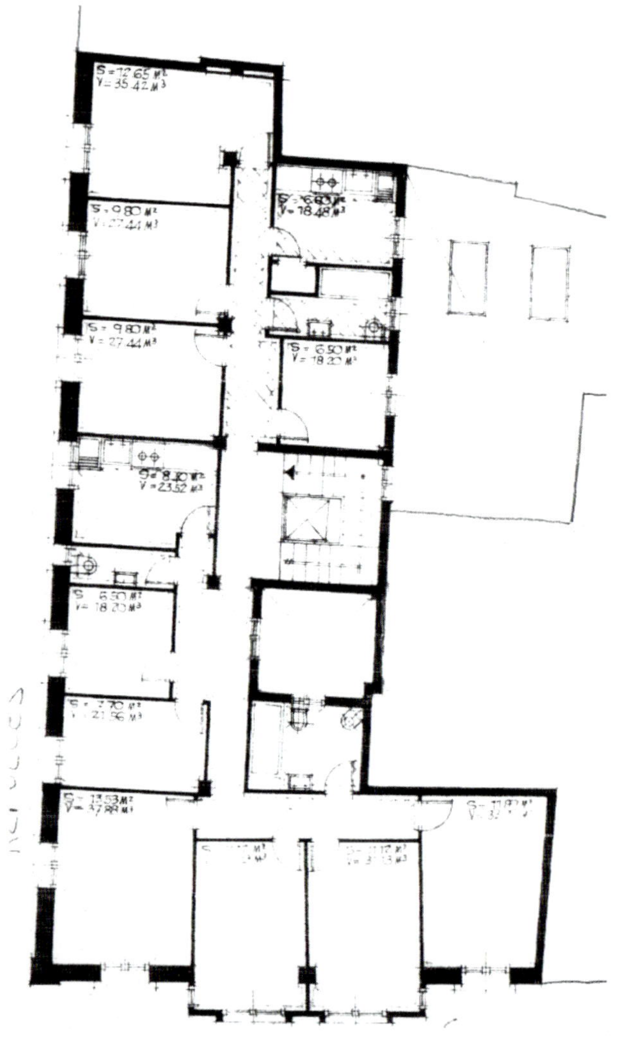

Existing floor plan

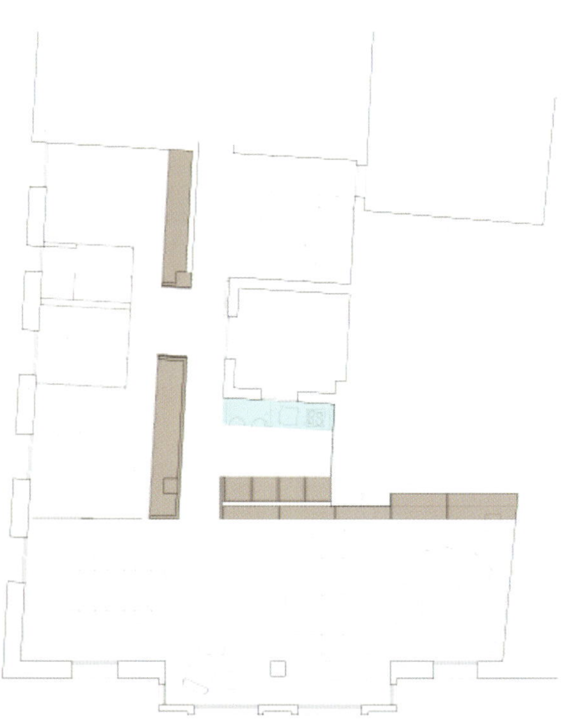

New floor plan

Sections

(Previous page). Photographs of
the reform. Inside the house, three
linear smoked oak containers define
the different areas, and a glass cube
houses the bathrooms. The spaces
are separated by sliding doors that
are secured in the ends of these
furniture-containers.

The architects paid special attention to the materials and finishes used in order to create interiors with restrained tones that provide a sensation of balance and serenity. For example, white oiled oak was used for flooring, adding harmony and elegance to the interior design scheme.

The furniture has been carefully selected to combine classic and contemporary styles. The Swan chairs by Arne Jacobsen have been combined with modern pieces such as the Metropolitan chair, the Tom Dixon S-chair and the Fiero Lissoni modular sofas. The dining room furniture consists of an Athos table by B&B Italia and Lia chairs by Zanotta.

Apollo Architects & Associates

> Tokyo, Japan | 2006 | Duration of project: 1 year | 926 sq ft | © Masao Nishikawa <

The client who requested the renovation of this apartment works as a camerawoman in the film and television world. This mother-of-two wanted the remodeling project to reflect her lifestyle, characterized by modernity and urban living. In addition to containing a comfy space where she and her children and live, she also needed a workspace.

Before the works, the space was gloomy, cold and not very practical. The design created by the architects is based on originality and minimalism, seeking tranquility, functionality and well-being. Part of the result of this process was the installation of dark wood flooring and blinds with vertical slats. These two design features give the dwelling a modern ambience.

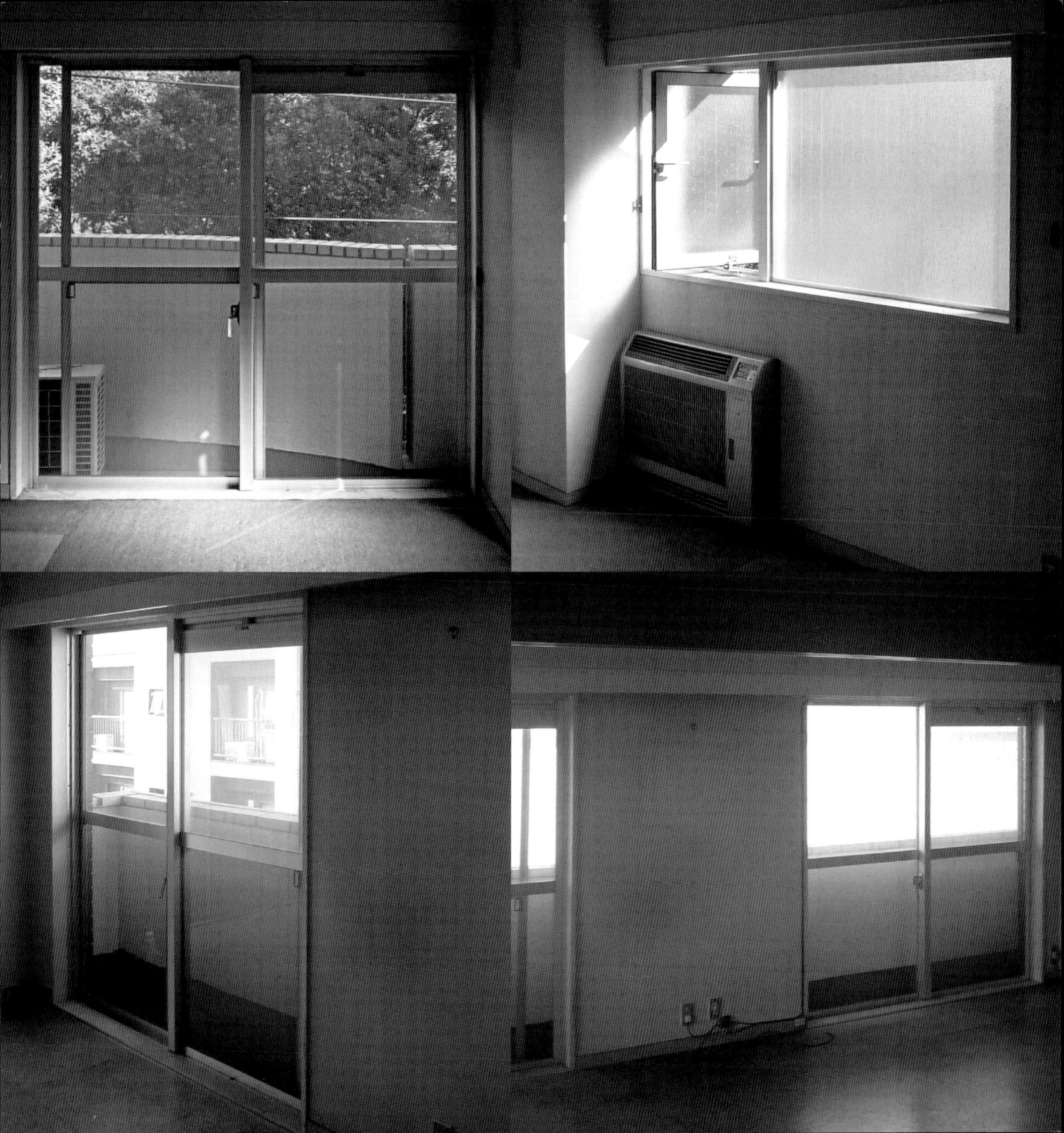

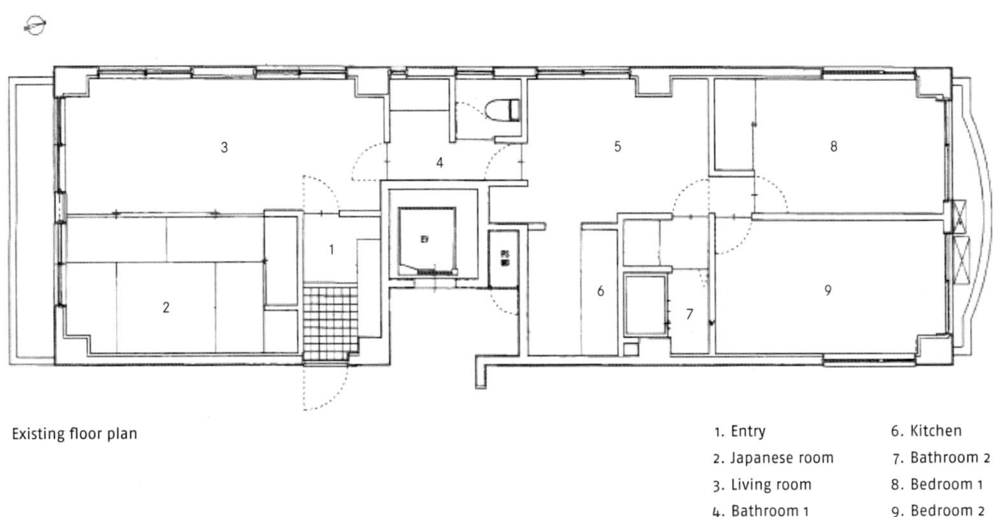

Existing floor plan

1. Entry
2. Japanese room
3. Living room
4. Bathroom 1
5. Dining room
6. Kitchen
7. Bathroom 2
8. Bedroom 1
9. Bedroom 2

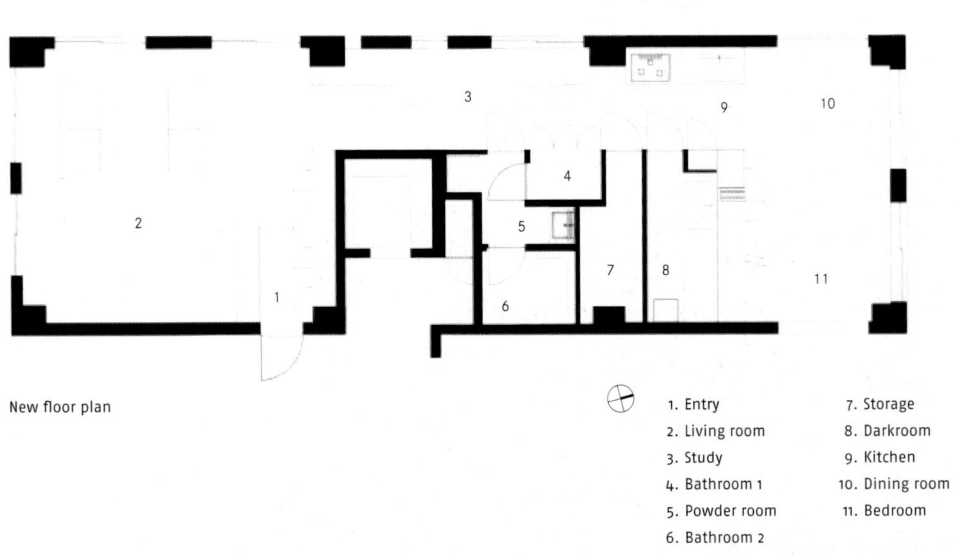

New floor plan

1. Entry
2. Living room
3. Study
4. Bathroom 1
5. Powder room
6. Bathroom 2
7. Storage
8. Darkroom
9. Kitchen
10. Dining room
11. Bedroom

This apartment is distributed around a central corridor that leads onto the different rooms. This layout favored the creation of a study against one of its walls. The study furniture consists of a bespoke elongated work desk and drawers, all made from wood. Two-tone pale colors prevail in the design, contrasting with the dark flooring.

The originality of this project is reflected in its furnishings and lighting. The architects designed a minimalist space where the furniture is sparse and the lighting is a result of a careful study. Artificial light emerges from small spotlights inset in the ceiling and distributed lineally along the corridor. Lights are also inset in the furniture, creating the effect that it emerging from these pieces. This type of lighting combines perfectly with the daylighting that enters through the large windows located in the lateral areas. This light is regulated by blinds with vertical slats.

Enrique Browne

> Las Condes, Santiago de Chile, Chile | 2008 | Duration of project: 8 months | 3,165 sq ft | © Felipe Fontecilla <

This makeover was carried out on an apartment that was built approximately 20 years ago. The dwelling, located in a building with some historical features, is located on a sixth floor, affording it excellent views to the west.

The dwelling's original program was remodeled to include a double bedrooms and a guest room. To do this, integrated, more ample spaces needed to be created. However, some of the apartment's features hampered the work: not being able to alter the building's exterior, its reinforced concrete walls and low ceilings.

The main changes were joining the living and dining rooms and enlarging the bedrooms. The spaces are unified by strips of wood installed in the roof.

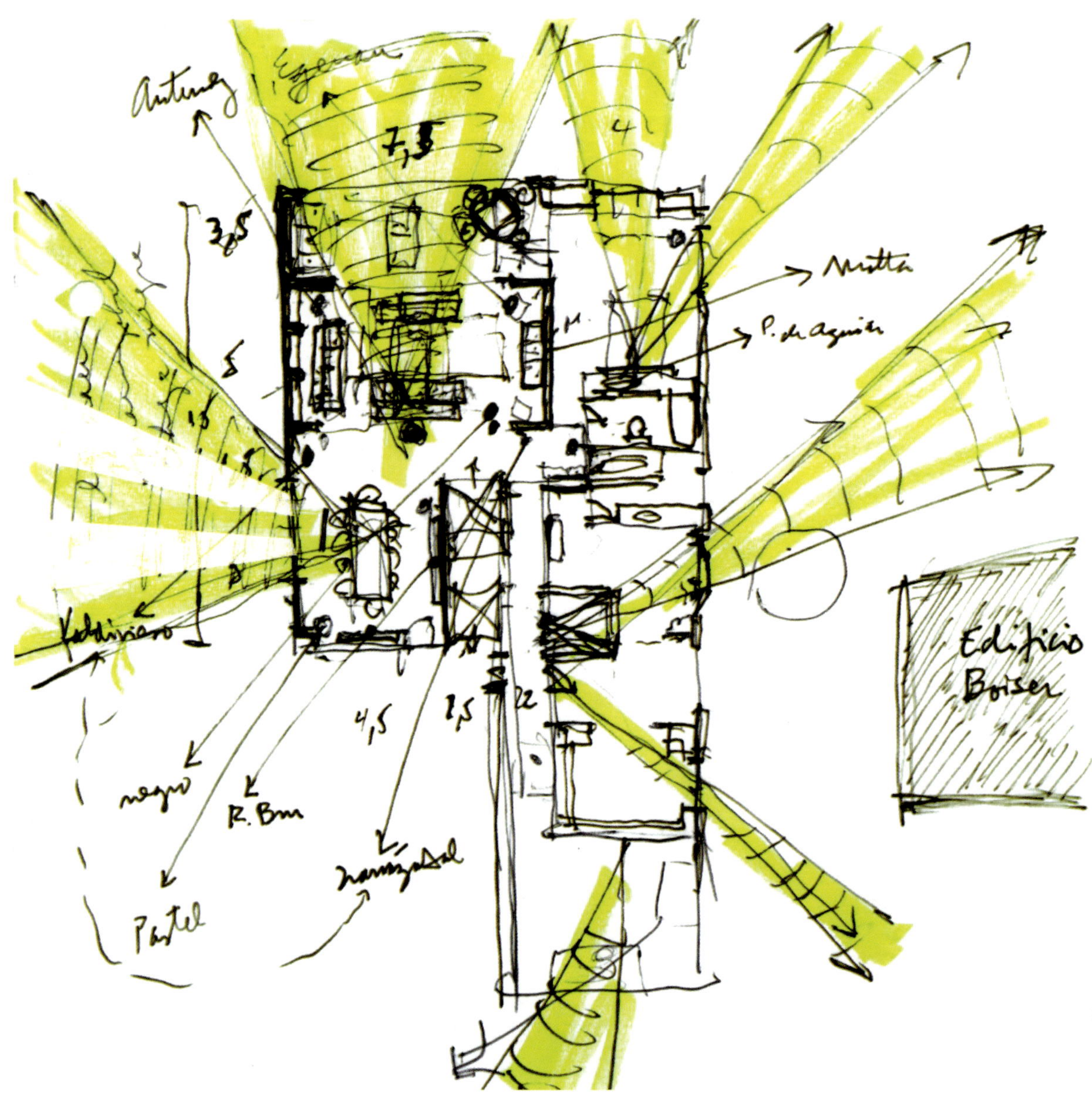

Preliminary sketch

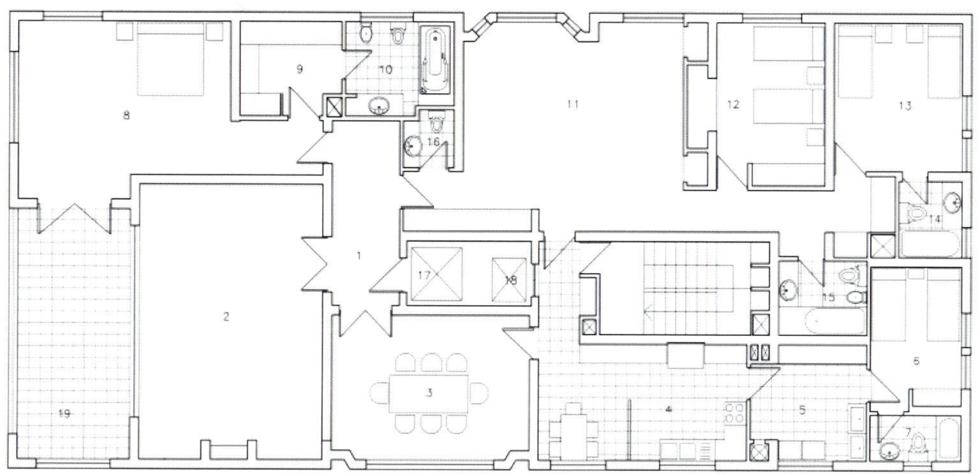

Existing floor plan

1. Entry hall
2. Living room
3. Dining room
4. Kitchen
5. Laundry room
6. Maid's room
7. Maid's bathroom
8. Master bedroom
9. Dressing room
10. Bathroom 1
11. Master living room
12. Bedroom 1
13. Bedroom 2
14. Bathroom 2
15. Bathroom 3
16. Bathroom 4
17. Master elevator
18. Maid's elevator
19. Terrace

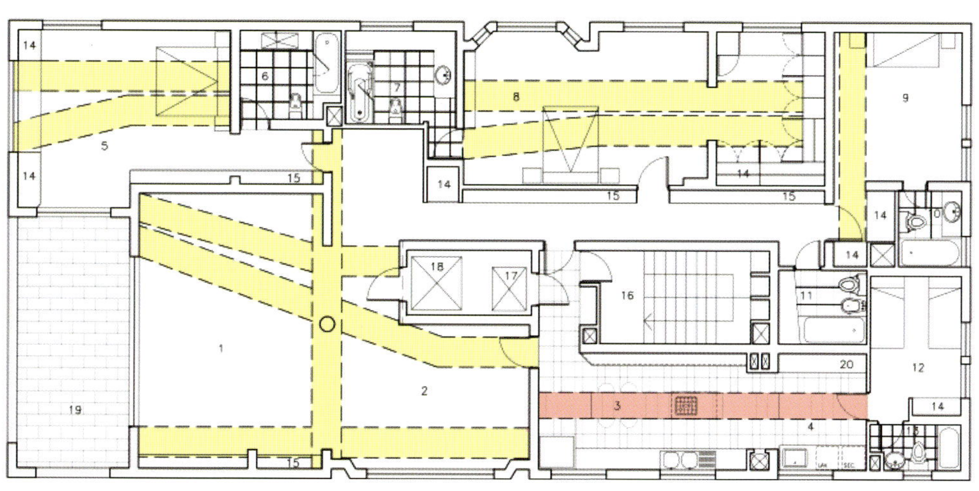

New floor plan

1. Living room
2. Dining room
3. Kitchen
4. Laundry room
5. Bedroom 1
6. Bathroom 1
7. Bathroom 2
8. Bedroom 2
9. Bedroom 3
10. Bathroom 3
11. Bathroom 4
12. Maid's bedroom
13. Maid's bathroom
14. Closet
15. Bookshelves
16. Fire escape
17. Maid's elevator
18. Master elevator
19. Terrace
20. Pantry

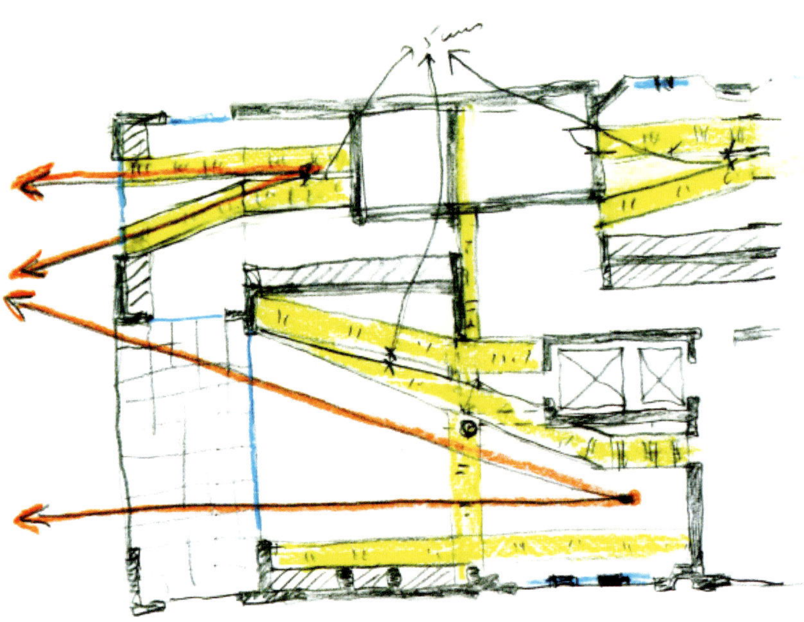

Study for the reflected ceiling plan

To counteract the handicaps of the original house –the reinforced concrete structure and the low ceilings– thin wooden strips were fitted in the ceilings, assisting the light fittings and other installations.

The thin wood beams installed in the ceiling run down to the floor in some doorways, thus visually and materially connecting the roofs and the floors. Parquet flooring was chosen to further unifying the space.

The architects played with the two types of light available. The combination of natural and artificial light gives variety to spaces such as the bathrooms, where the wood was spatially arranged to generate maximum warmth.

In the kitchen, red ceramic strips were created, which run from the floor to the ceiling. The route continues over the dining room and then down again and across the floor. This system seeks continuity, obtained by using a single color and material.

ALUMINUM HOUSE

> Yokohama, Japan | 2008 | Duration of project: 5 months | 506 sq ft | © Archipro Architects <

The project was to renovate an apartment located in the heart of the Japanese city of Yokohama. The prospective tenants, a young couple of newlyweds, wanted to live in the city center to be close to leisure facilities and their places of work.

The block of houses was originally an office building. The clients wanted to transform the office into a dwelling that would function as a home and an office at the same time.

The result was a simple but comfortable space, with just a single bedroom. Very few changes were made to the interior layout, only a few layers were added to the original walls. The unique nature of the project stems from its roof constructed from L-shaped aluminum channels, which are aligned as if they were floating. This architectural solution gives the house its name.

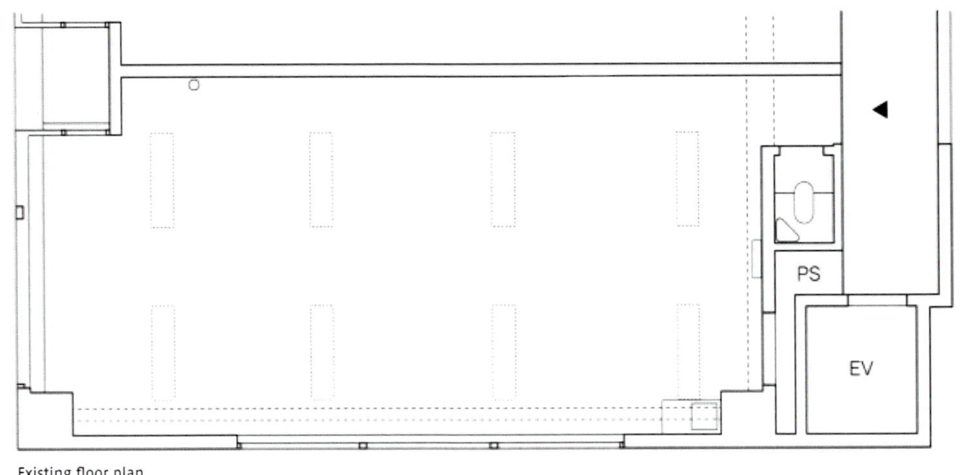

Existing floor plan

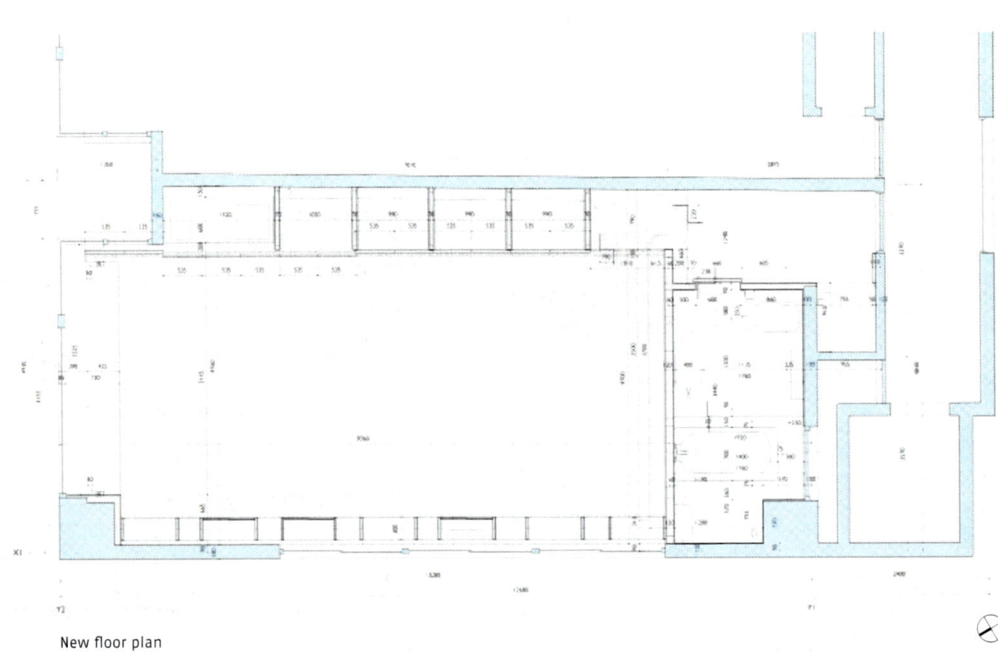

New floor plan

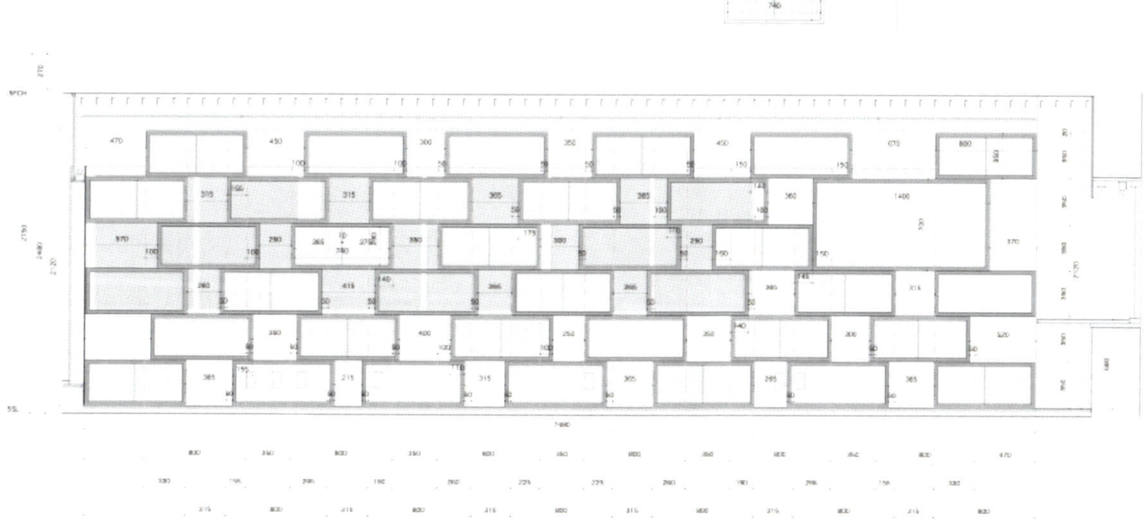

Bookshelf elevation

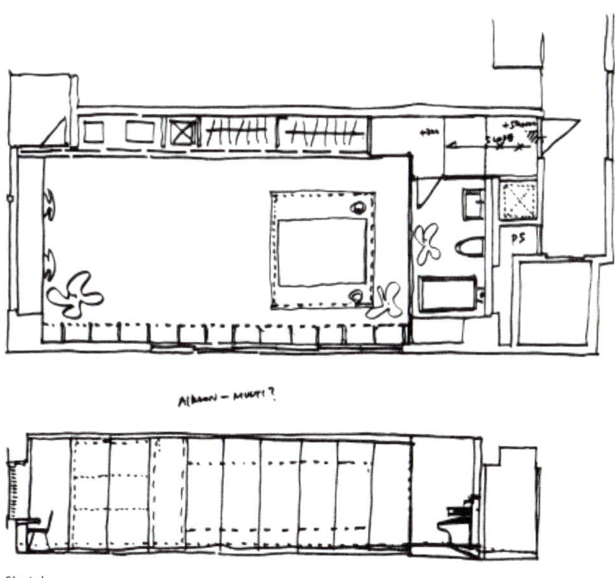

Sketch

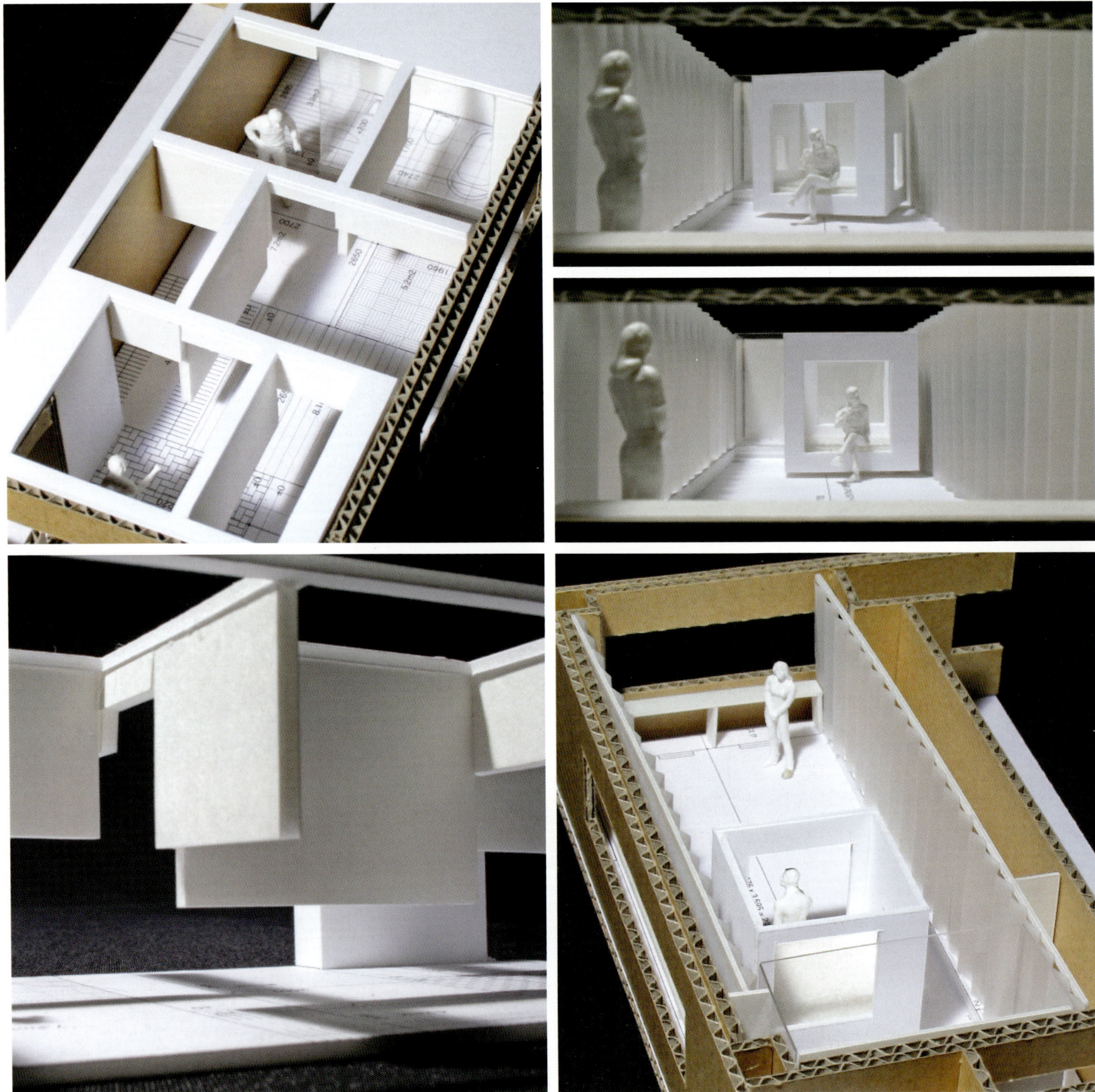

The architect's main objective was to create a home that was both an architectural space and a showcase for lighting accessories. To achieve this, the architect took advantage of the existing walls and inserted thin surfaces into them to define the different areas. One of the key features of the project is the roof, which has been installed over the former one and constructed from small L-shaped sheets of aluminum.

Against one of the existing walls, a large shelf was built from small white boxes of different shapes and sizes, distributed along the length and width of the wall. This lets in natural light, filtered by translucent panels, and can be used to store all kinds of objects, from books to shoes and other accessories.

The wardrobes and kitchen are located behind narrow sliding doors installed in the existed partition wall. For the flooring, the architects designed a thin layer of pine wood with small spoon-shaped cuts in the surface. The master bedroom faces north so it receives little sunlight.

Paolo Cesaretti

PRIVATE HOUSE IN FLORENCE

> Bagno a Ripoli, Italy | 2005 | Duration of project: 2 years | 1,130 sq ft | © Paolo Cesaretti <

The house is situated on the top floor of a 17th century building located in Chianti, Tuscany, near Florence. Originally, the building functioned as a covered pavilion for *pallacorda* a typical Italian game, considered to be a forerunner of fronton. The construction was part of the architectural complex of Villa Medici di Lappeggi.

The main objective of the remodeling was to restore the structure and then transform it into a house. The main concern of the architects' studio was to materially and structurally conserve the old structure and retain the original sense of space.

The countryside surrounding the house is framed by six large windows, one of the key features of the house. This landscape, so typical of this area of Italy, had to be fully captured in a single glance. The project was therefore based on the decision to preserve the interior perimeter of the house, without any extra additions.

BEFORE

AFTER

Existing floor plan

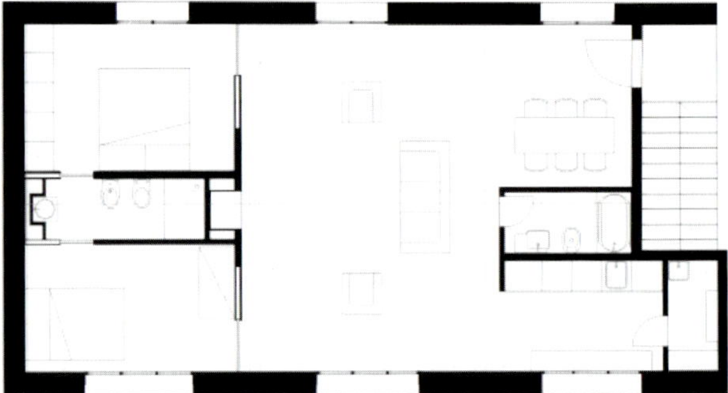

New floor plan

The design created by the architects had to integrate the new features into existing ones in order to maintain their separate identities. The result is a house that balances a city lifestyle with the ways of the countryside.

The materials, colors and surface treatments, as well as the lighting effects, were chosen to emphasize the connection to the historical architectural context. The local stone used in walls and timber beams, representing the old material, combine perfectly with the parquet floor and wall tiles in the bathroom, representing the new.

RVR Arquitectos

HOUSE IN PORTA DA PENA

> Santiago de Compostela, Spain | 2007 | Duration of project: 6 months | 1,023 sq ft | © Héctor Fernández Santos-Díez <

This little dwelling is located in the old town of Santiago de Compostela. It consists of an apartment that was remodeled in the early 1950s, when the garret was converted into an attic. The result was a chaotic framework, with multiple beam supports and intermediary supports on the thin walls that supported the roof.

The remodeling criteria was based on the well-known phrase that Don Fabrizio Corbera, Prince of Salina, says to Aimone Chevalley in the famous novel *Il Gattopardo*, written by Giuseppe Tomasi di Lampedusa: "If we want things to stay as they are, things will have to change". The original materials, roof and façade openings were maintained and a coherent structural design was applied. The layout of the kitchen, bathroom, windows, and entrance were changed.

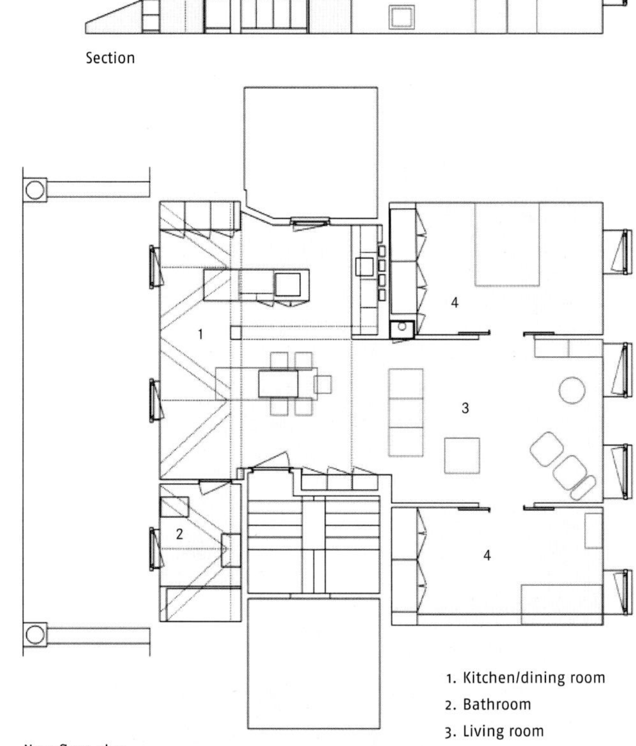

Section

New floor plan

1. Kitchen/dining room
2. Bathroom
3. Living room
4. Bedroom

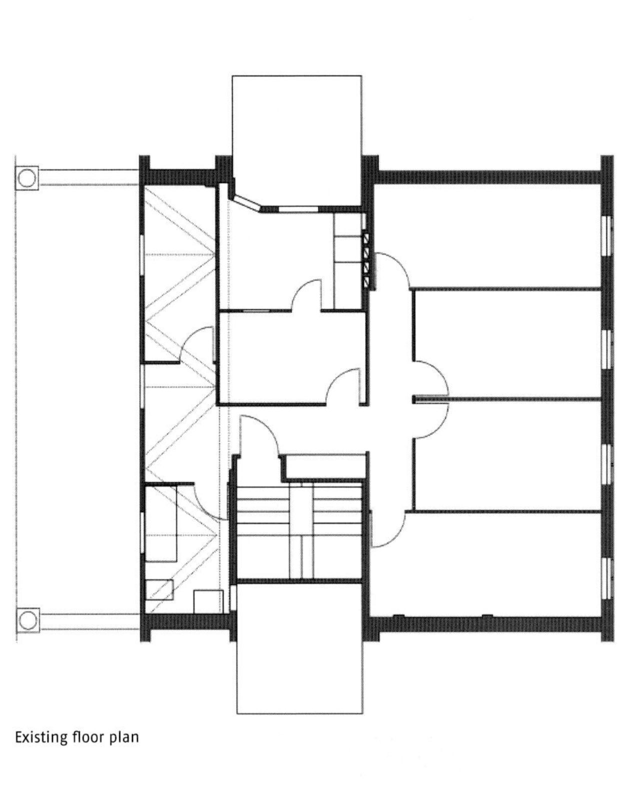

Existing floor plan

0　　　5m

In the interior layout, two spaces were created as through they were boxes. Large sliding doors were installed and, depending on the owners' priorities, these can be opened to integrate these spaces into the common area, or closed to provide privacy.

The new structure is based on acknowledging greater or lesser privacy, depending on the activities carried out in the different spaces. For example, the kitchen is connected to the living area, the bathroom is isolated from the rest of the rooms, and the two complementary rooms can adapt to different uses, depending on the occupants' needs.

Plystudio

> Singapore, Singapore | 2008 | Duration of project: 5 months | 1,292 sq ft | © Stzernstudio <

This typical post-war attic was remodeled to meet the client's particular needs. All the inner walls were knocked through to create an ample space that enables larger quantities of daylighting to enter. The end result is an open-plan space that fulfils the multiple functions of living space, dining area, study and bedroom with bathroom.

The architects came up with different design strategies to define the different spaces. In the end, they opted for a series of wood boxes in a linear arrangement, similar to library shelves. These structures enable the different areas to function independently, while all forming part of the central space. The structures only take up 6.5 ft of the apartment's 20 ft width and are distributed longitudinally. The only enclosing structures are the doors used to close off the private areas.

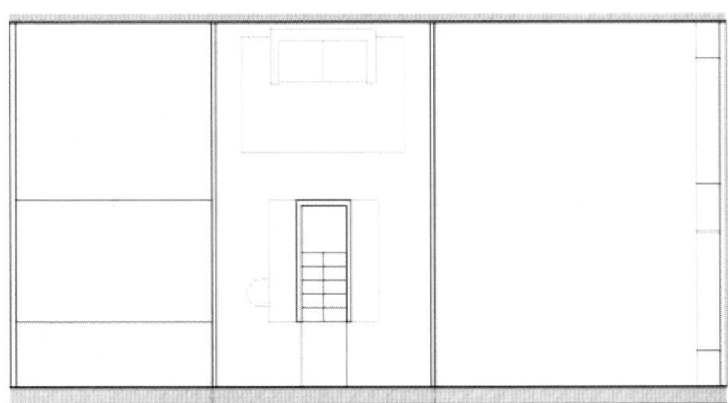

Mezzanine plan

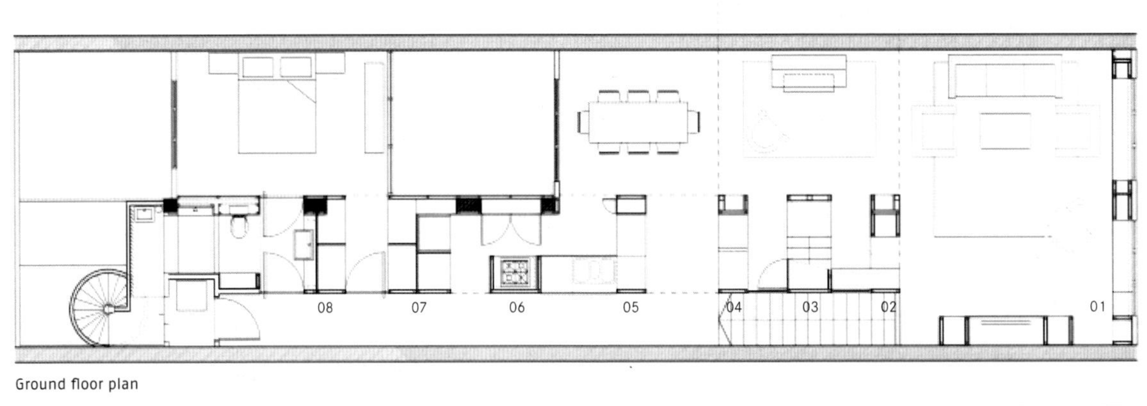

Ground floor plan

0   5m

 01
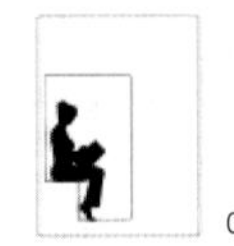 02
 03
 04
 05
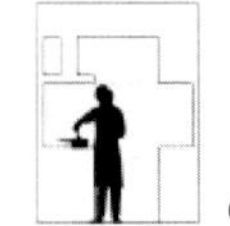 06
 07
 08
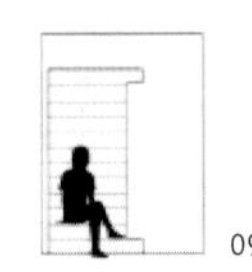 09

Diagrams

(Previous page). Diagrams showing the different wood box models that define the loft's different areas. The spaces created include a chill-out seating area (1), a reading area (2), stairs to the open gallery (3), and the kitchen (6).

Knocking down the walls separating the different rooms enabled more daylighting to illuminate the 1290 sq ft attic. Each end of the house contains doors and windows, and small spotlights were fitted to create an interplay of lighting and shading.
The structures are made from 1/2" plywood. In this project, the architects focused on creating innovative details, treading a fine line between architecture and interior design.

RESIDENTIAL CONTAINERS

> Prague, Czech Republic | 2008 | Duration of project: 2 years | 1,625 sq ft | © Esther Havlová <

Petr Hájek/HŠH Architekti

This original project consisted in adding two rectangular modules onto the roof of a building in downtown Prague. Extensions should not affect the legibility of the original building. These ones were designed to accommodate spaces that complement the rest of rooms in the attic.

These containers house living spaces, bathrooms and service installations. The other rooms and the wardrobes were located in the original building. The structure of these modules allows folding walls to be inserted in order to define separate spaces. Openings have been made in the outer walls of the containers to insert big windows that allow light to enter and the interiors to be visible from outside.

All the structures have been built using technologies commonly employed in construction and basic materials such as wood, glass, metal and drywall.

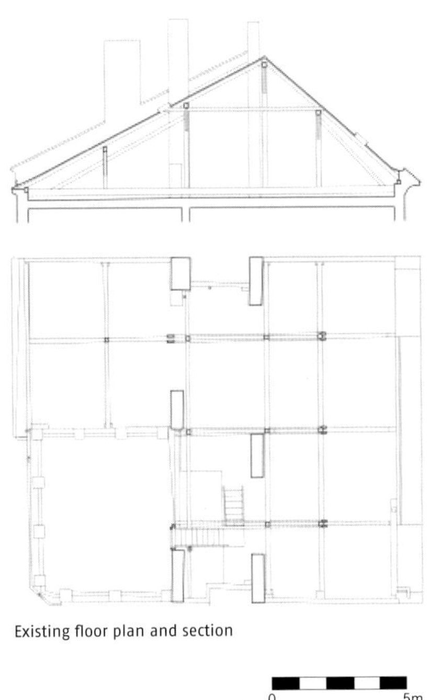

Existing floor plan and section

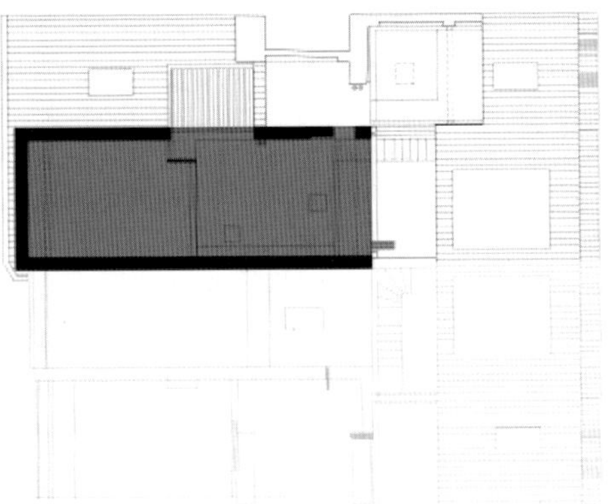

New attic upper level

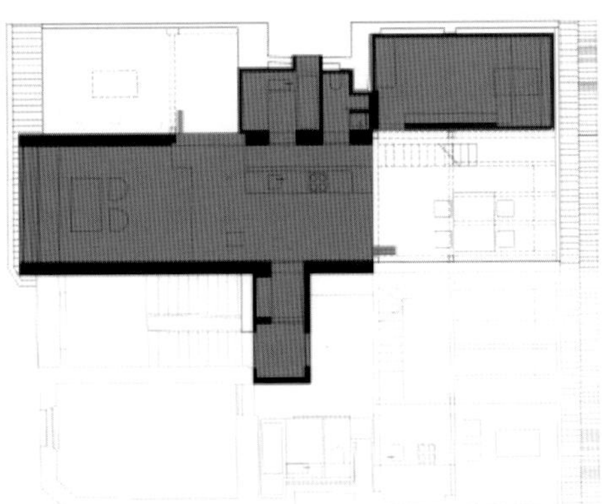

New attic lower level

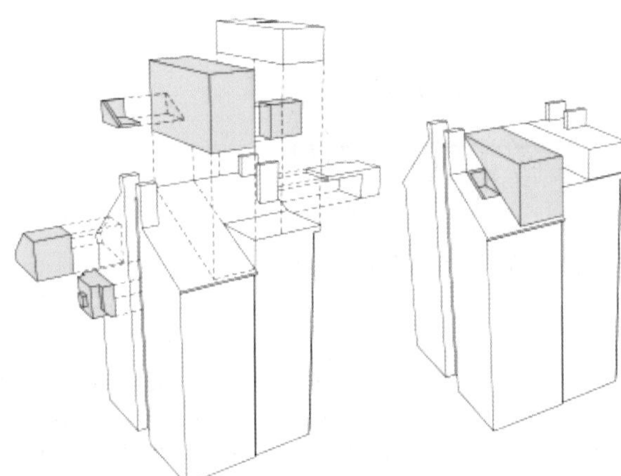

Exploded axonometric

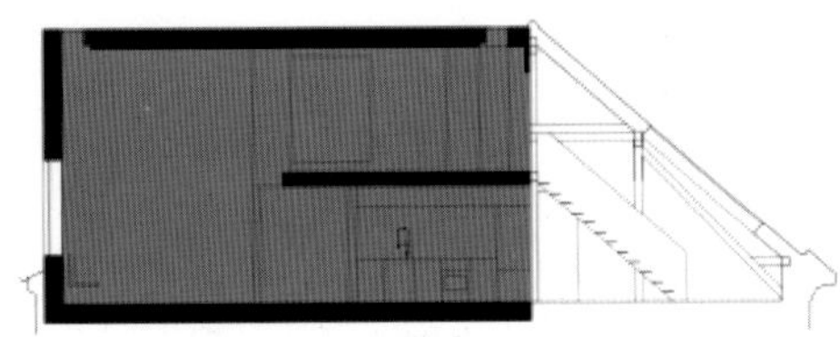

Section

Folding walls were installed that serve as space dividers or sliding doors that provide the privacy needed by the occupants. These created compact volumes that can be used to isolate spaces.

The new features in this intervention have either been covered with small titan-zinc sheets, or painted gray. There is a continuous mix of materials typically used in construction such as wood, metal, glass and drywall. The more public spaces, such as the kitchen, living room and dining room, have been located in these containers.

L RESIDENCE

> Omaha, NE, USA | 2008 | Duration of project: 18 months | 3,400 sq ft | © Larry Gawel <

This apartment, designed for a film maker, is a reinterpretation of *poche* spaces, forming an architecture of sequences and different perspectives. In this design, the virtual *poche*, which the Baroque style used to conceal washrooms or create mysterious and surprising spaces, takes on a film theme that places more emphasis on surface than on solidity or mass.

The dwelling is situated in a former art deco style hotel that has been recently converted into apartments with ground floor commercial premises. Although the rooms have high roofs, the attic windows have the same proportions as the windows in the lower levels, reaching a height of 7.4 ft. The project proposal focused on reorganizing the main space along the perimeter close to the openings to the exterior, and in concealing all utility areas behind an oak-veneered wall.

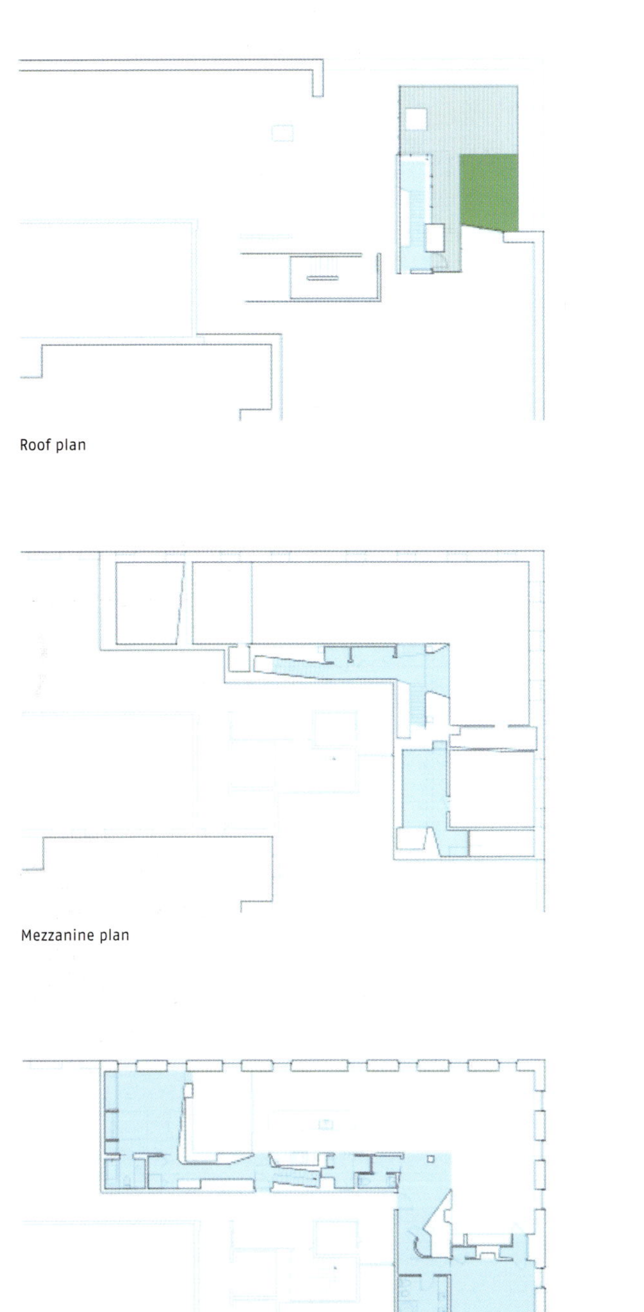

Roof plan

Mezzanine plan

Main floor plan

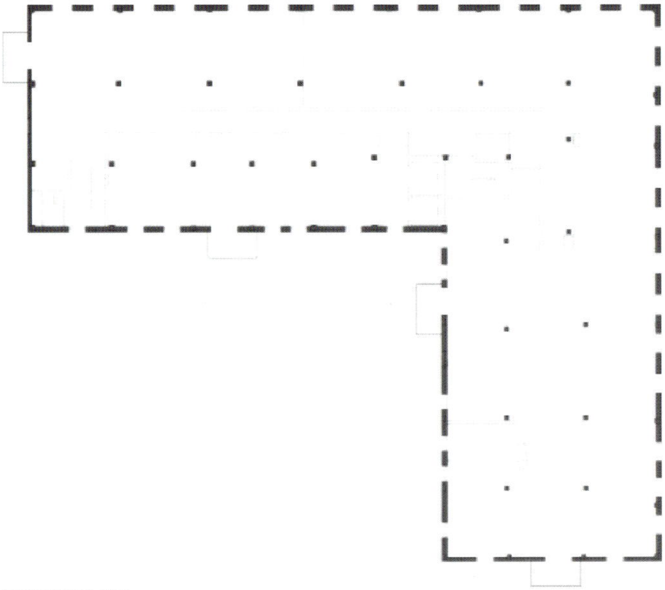

Existing floor plan

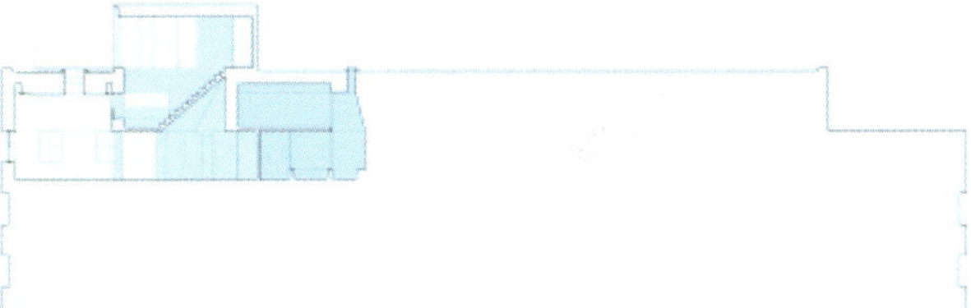

North–South section

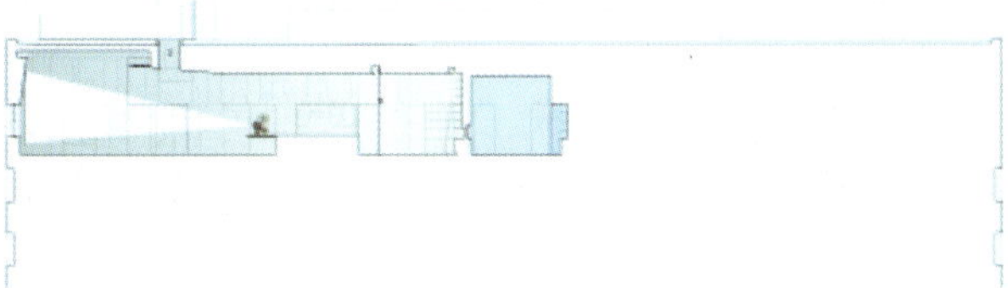

East–West section 1

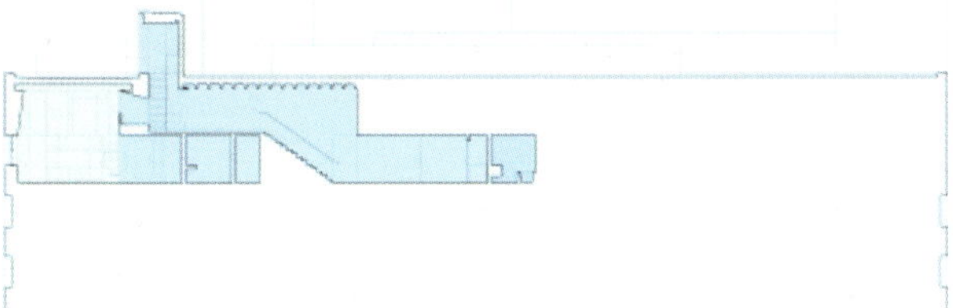

East–West section 2

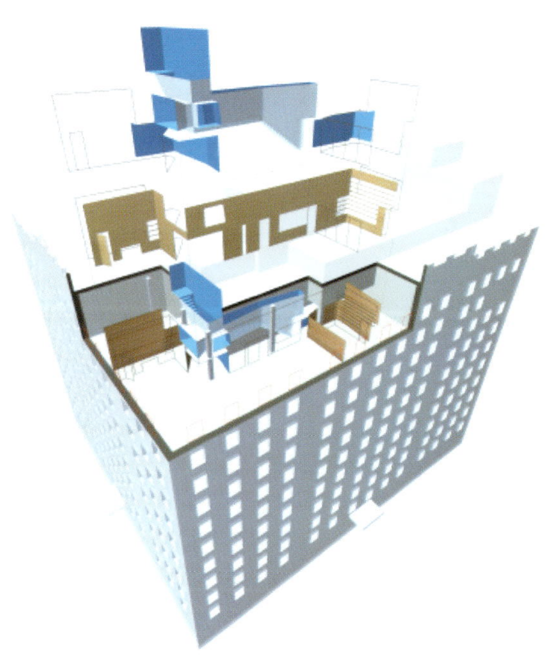

Surfaces Diagram

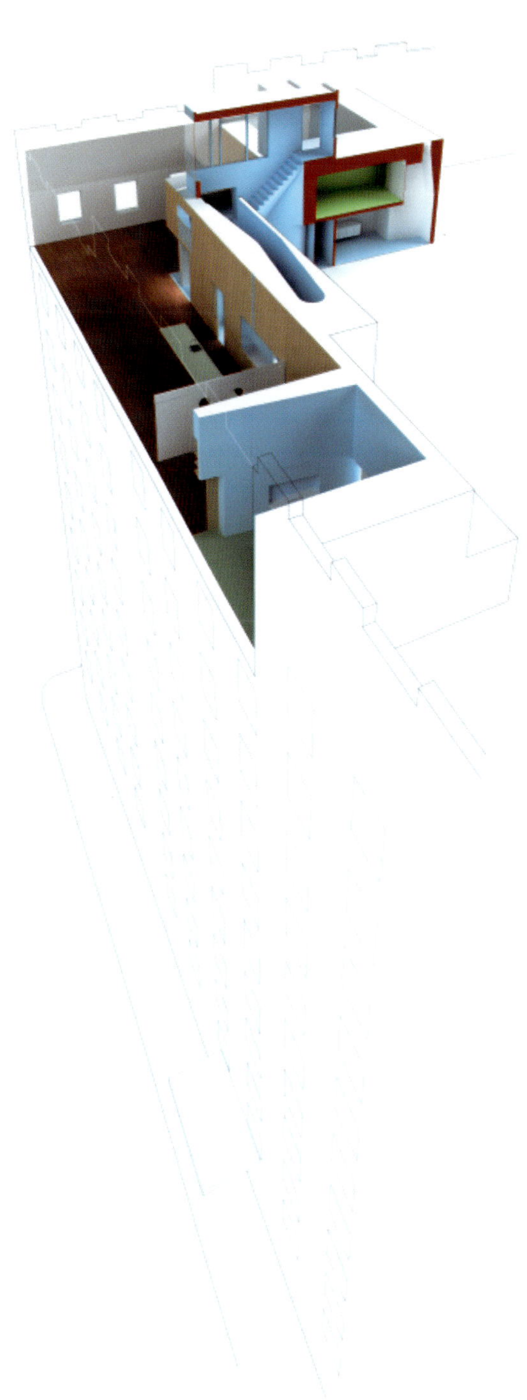

West section perspective

The strategy behind the *poche* spaces becomes clear when looking at the floor plan organization: the private or secondary spaces, such as the bathrooms or storage areas, are separated from the main space by an L-shaped wall. The bedrooms are situated in either end of the *poche* space to make the most of the daylighting.

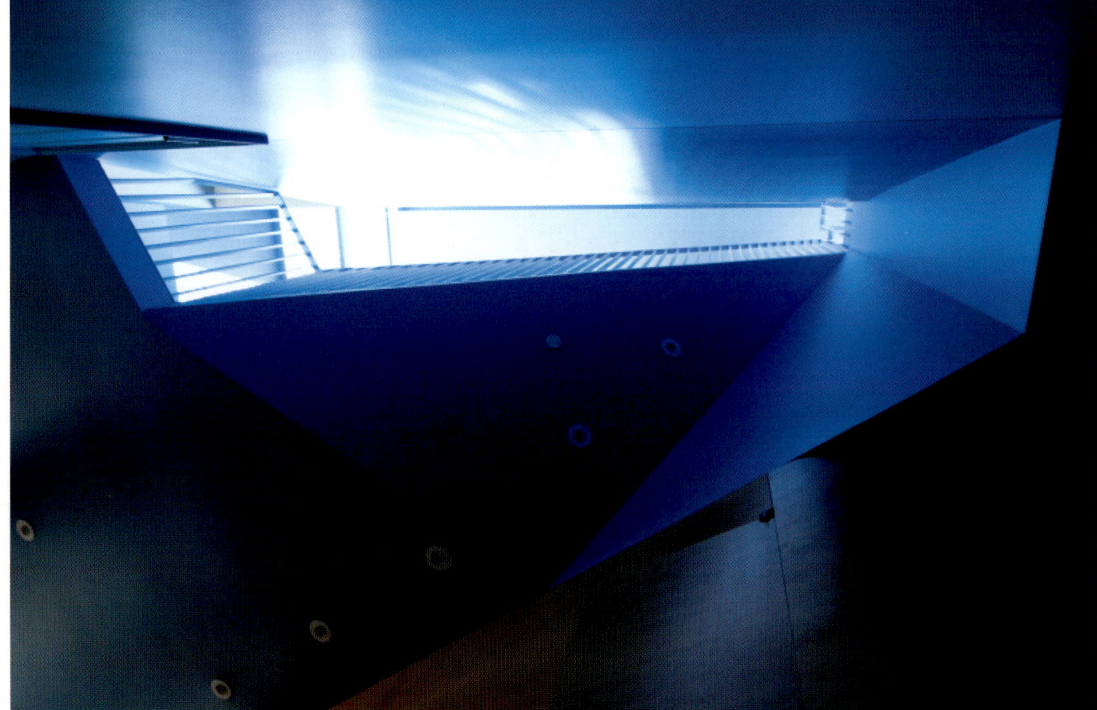

A large part of the *poche* space was painted blue to enhance the wood surface and create a contrasting effect. Different colors were chosen in the bedrooms and bathrooms to give them character.

Paul Cha Architect

CHELSEA DUPLEX PENTHOUSE

> Chelsea, NY, USA | 2008 | Duration of project: 10 months | 2,850 sq ft | © Dao Lou Zha <

This loft is situated in the uppermost story of a glass factory that has been recently remodeled into a block of apartments. A small attic was also added during the conversion. The client, a young investments broker, commissioned a space for living and another for working, suitable for both working and entertaining friends.

This early 20th century building has features that are typical in the Chelsea district: a rectangular floor template, exterior brick walls, and oak parquet over pine floor boards. The attic, which was added at a later date, has a sloped roof inset with skylights. The renovation strategy consisted in inserting architectural features inside the existing structure. The loft is divided into three areas: a living area, a kitchen-dining area, and the bedroom-bathroom space. The work zone is located in the attic, above the kitchen-dining area.

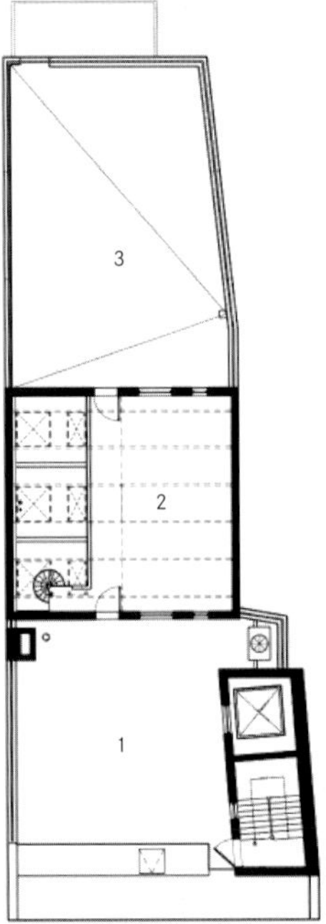

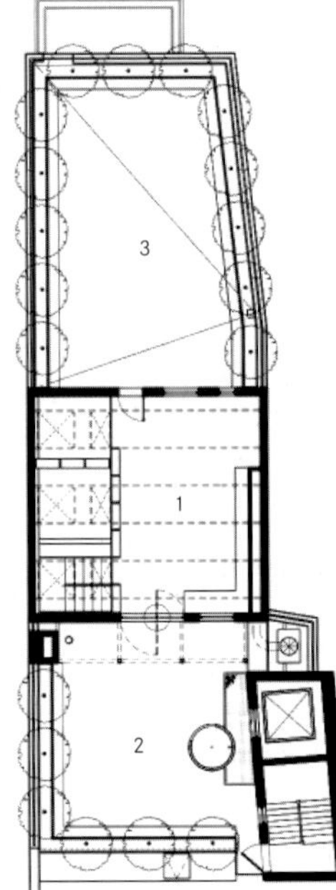

A series of walls and shelves fulfill occupants' functional requirements. Sliding and pivoting doors create a sensation of amplitude. The carefully thought out layout creates a pleasant and unifying space.

Existing penthouse plan

1. South roof terrace
2. Roof study
3. North roof terrace

New penthouse plan

1. Study
2. South terrace
3. North terrace

Existing lower level plan

1. Elevator
2. Fire stairs
3. Stairs
4. Closet 1
5. Entry
6. Living room
7. Powder room
8. Kitchen
9. Dining room
10. Closet 2
11. Master bathroom
12. Master bedroom
13. Bedroom
14. Balcony

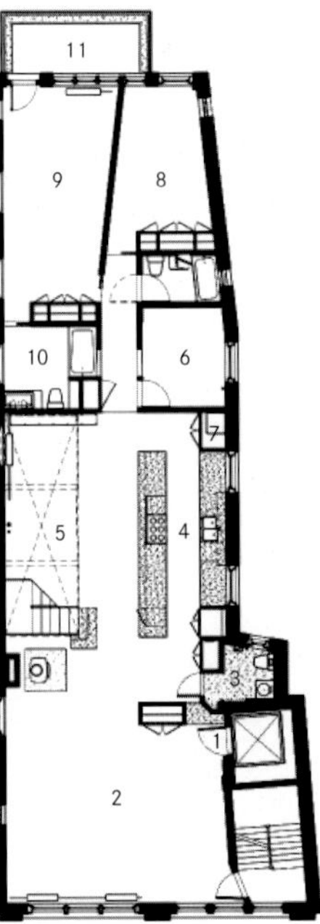

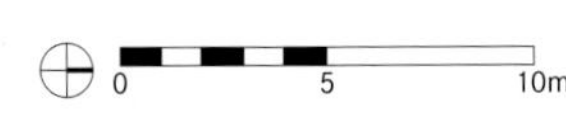

New lower level plan

1. Entry
2. Living room
3. Powder room
4. Kitchen
5. Dining room
6. Guest room
7. Bathroom
8. Bedroom
9. Master bedroom
10. Master bathroom
11. Balcony

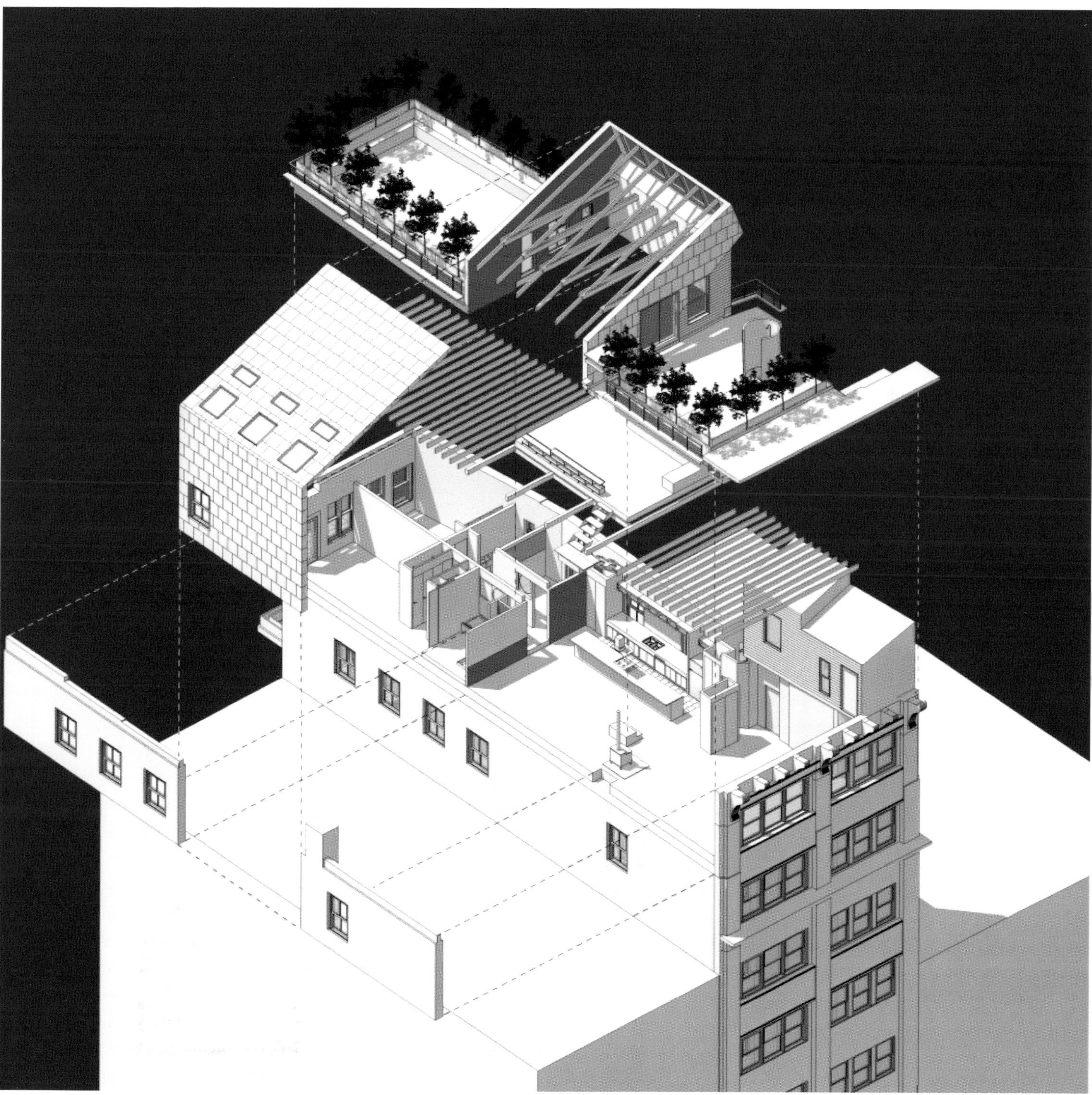

Exploded axonometric

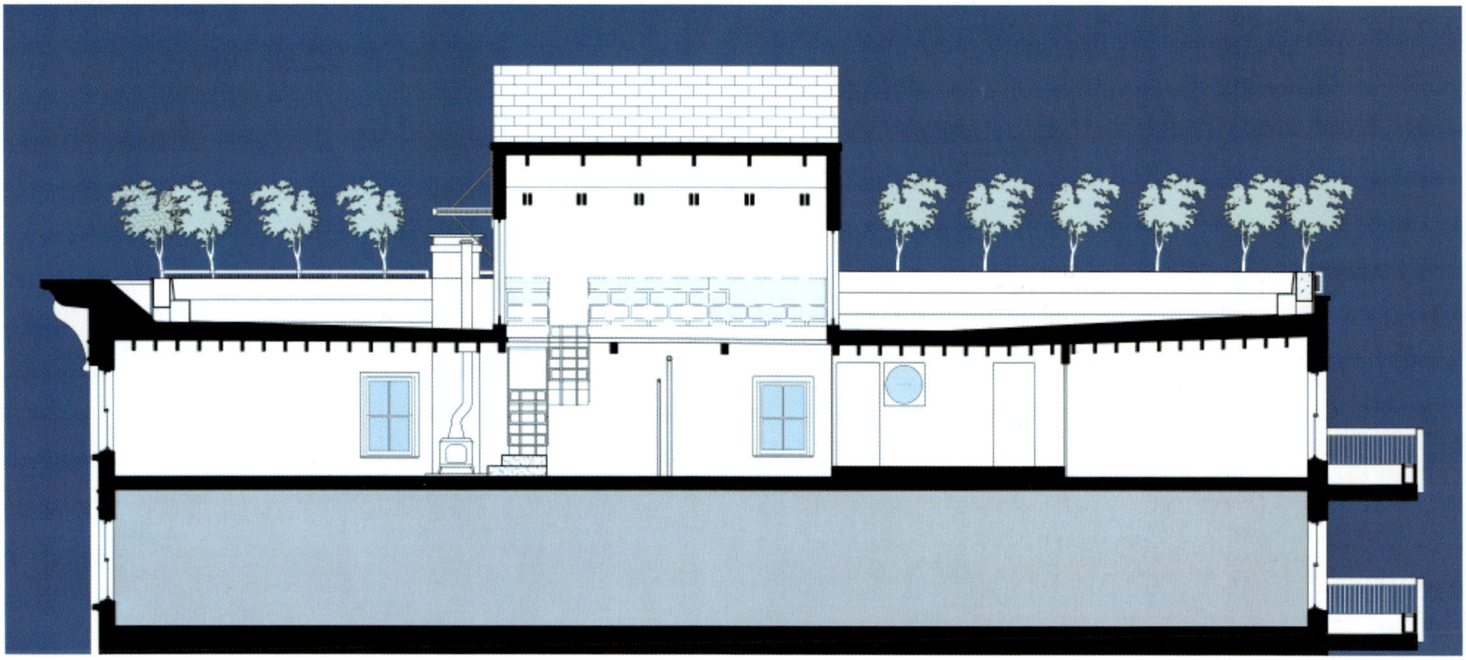

Longitudinal section

The design strategy is unified through finishes and colors. All the existing finishes were conserved to form a contrast with the new architectural features, characterized by meticulous forms, colors and textures.

The storage spaces are based on a module that is repeated throughout the dwelling and which adopts different forms to fulfill different functions. For example, the storage space in the base of the kitchen island is echoed in the kitchen cupboards, and it reappears in the same shape, yet deconstructed, in the stair treads.

In the attic, the office has access to the north- and south-facing terraces; a staircase with a sloping roof inset with skylights connects this space to the level below. Storage elements allow the available space to be optimally used and create dividers without breaking the uniformity of the space.

LOSA LOFT

> San Francisco, CA, USA | 2007 | Duration of project: 11 months | 1,528 sq ft | © Matthew Millman <

Aidlin Darling Design

This apartment is located in the La Misión district in the Californian city of San Francisco. The clients, a young couple, wanted to transform this loft whose design did not suit their needs. They needed a highly function space, with clear details and which reflected a style characteristic of a way of urban living.

The remodeling was carried out in two phases using two different solutions. The first intervention consisted in cleaning the interior spaces, as if it were a white, neutral box. The second phase strategically inserted five architectural features that take center stage in the project: "hairpin", "zipper", "fireplace", "platform" and "screen". These give rhythm to the space, which has no staircase. The shape and connection of the features were directly designed in accordance with the clients' brief.

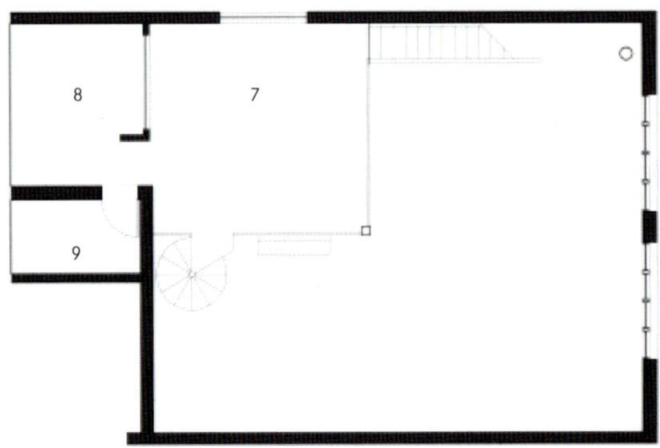

Existing mezzanine plan

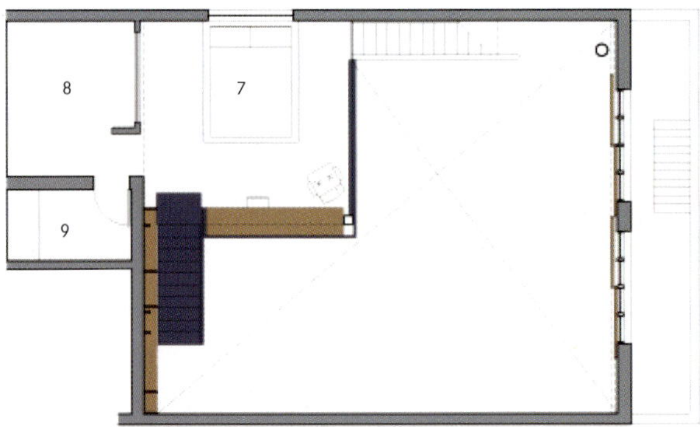

New mezzanine plan

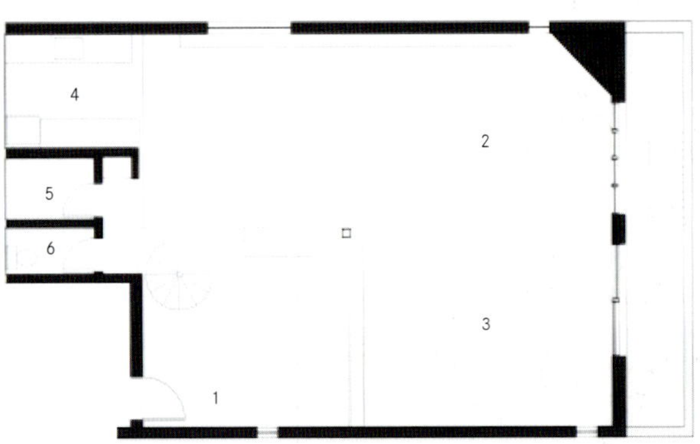

Existing ground floor plan

1. Entry
2. Dining room
3. Living room
4. Kitchen
5. Pantry
6. Powder room
7. Loft
8. Closet
9. Bathroom

New ground floor plan

1. Entry
2. Dining room
3. Living room
4. Kitchen
5. Pantry
6. Powder room
7. Loft
8. Closet
9. Bathroom

The "hairpin" and "zipper" are two of the five main features in the project. The first contains two work spaces, a book shelf, a wardrobe and a storage space, while the second contains a wood and steel staircase, the only element connecting the spaces located on different levels.

The three other architectural features house the rest of the spaces in the following arrangement: the "fireplace" area contains a storeroom and the fireplace; the "platform" houses the kitchen with central island and breakfast bar; finally the dining room contains a structure that functions as a screen, which provides insulation and security when the exterior doors are opened during the summer months.

CHIHUAHUA 78 HOUSES

> Mexico City, Mexico | 2006 | Duration of project: 1 year | 9,537 sq ft | © Luis Gordoa, Jair Navarrete <

This project is located in the Colonia Roma district in Mexico City. This area stems back to the colonial era when it was populated by the burgeoning middle and upper classes. This building was constructed in 1916 and was the work of the company Arquitectura Prunes.

The artistic style of the façade could be easily placed within the canons of art nouveau. This artistic and architectural style can be found in each of the elegant ornamental details of the houses and buildings in this part of the city. In this particular building, the interior is eclectic, typical of a time when architectural style focused on ornamental stone and plasterwork.

Over time, the building became worn, so the architects' intervention restored the façade and remodeled the interiors. A new three-level steel structure was created in the rear of the property.

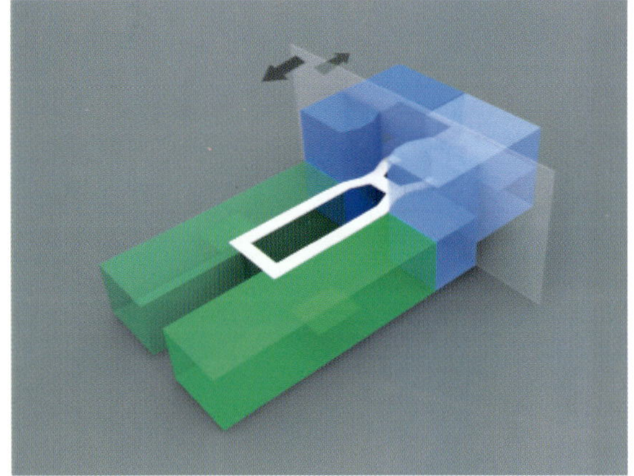

Space tipologies models 1

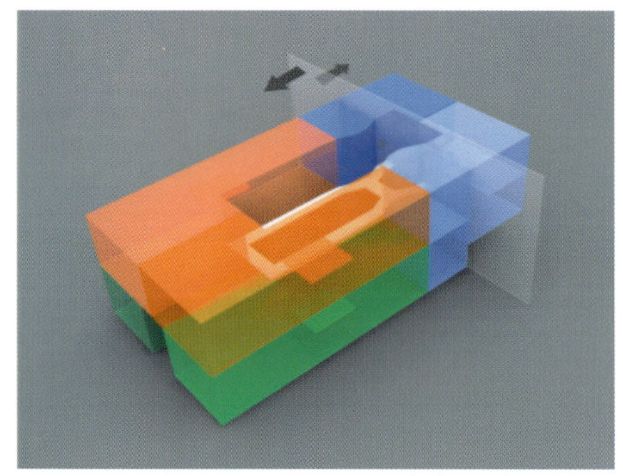

Space tipologies models 2

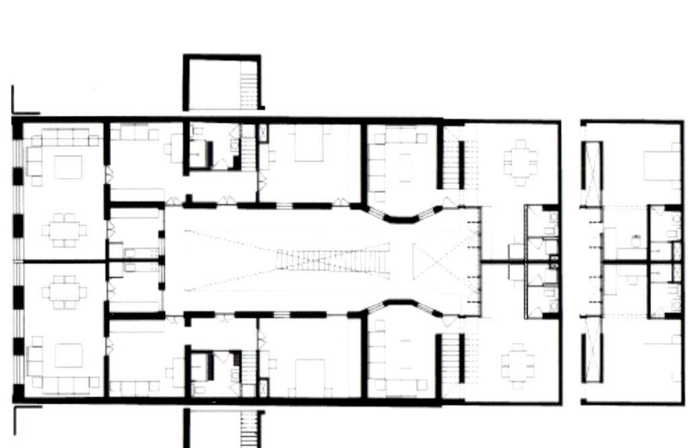

Upper level plan

Lower level plan

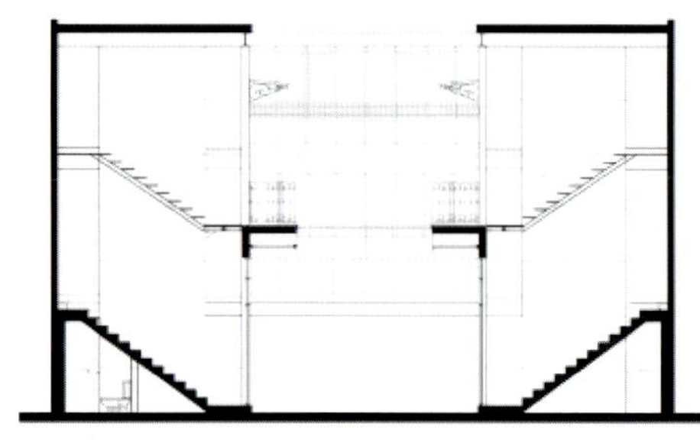

Cross section

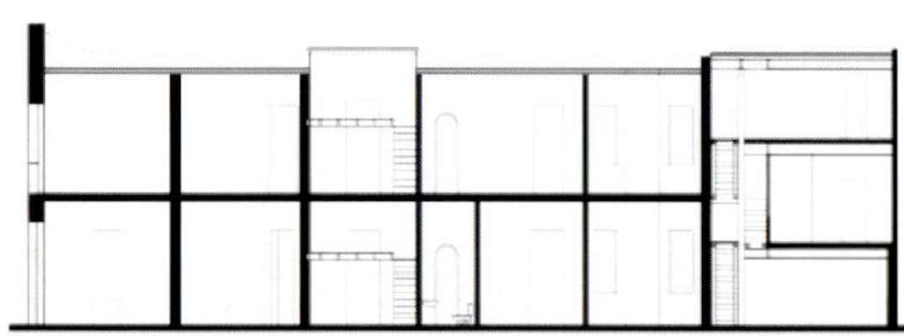

Longitudinal section 1

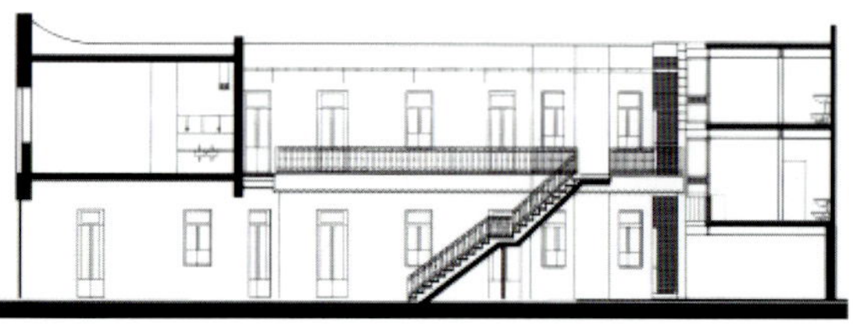

Longitudinal section 2

The two main floors of this mansion, containing six apartments, contain four half levels. The additions that were made in the remodeling allow this new half-level distribution of space. The upper level is accessed by the original main staircase, which has been restored.

While maintaining the original structure, the architects added details that give the building a contemporary character. The finishes, the paste mosaic floors, and colored seals in the wooden windows and doors all respect the artistic spirit of the original design.

Juan Diego López-Arquillo/LAA Arquitectos

SAN MATÍAS HOUSES

> Granada, Spain | 2006 | Duration of project: 11 months | 4,327 sq ft | Javier Algarra, Aurelio Dorronsoro <

In their early life, these 19th century houses, located in the old town of Granada, were bordellos, which were gradually converted into residences. The building's new owner commissioned a renovation project to create a restaurant at street level, and an office and two small apartments in the levels above.

The remodeling project consisted in restructuring the space and material used in the building's interiors. The architects enhanced both the transversal character of the corridor starting in the entranceway opening onto Calle Palacios, and the orthogonal shape of the upper entrance. The two corridors are used to connect the ground floor and the lower-ground floor. The vertical connections are made by means of adjoining staircases, located only on two levels, therefore giving the interior courtyard a central role in the house.

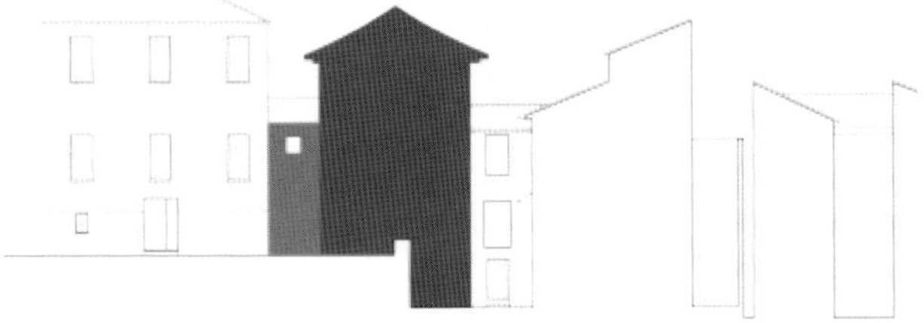

Existing condition

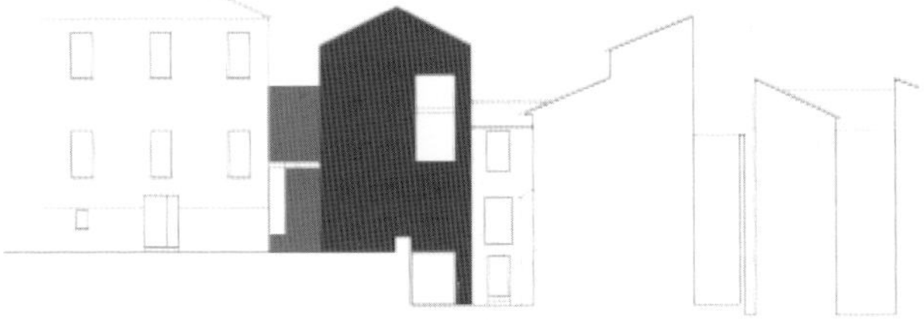

1999

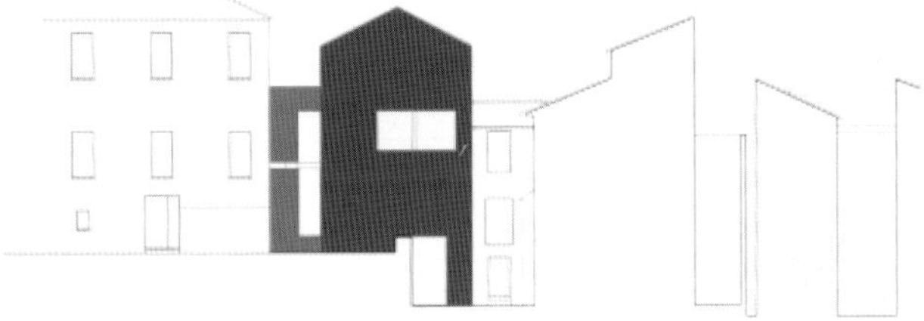

2001

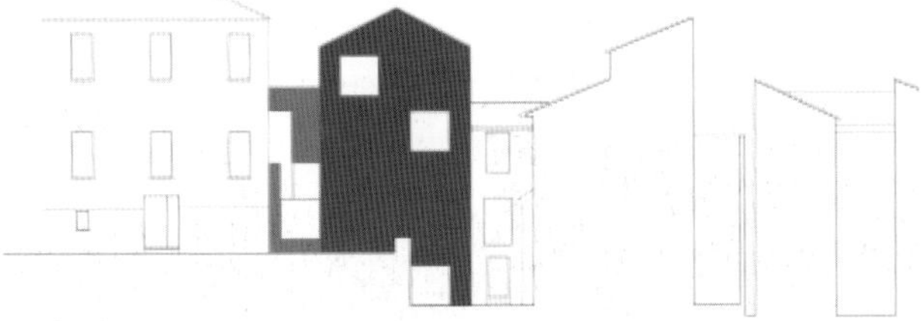

2003

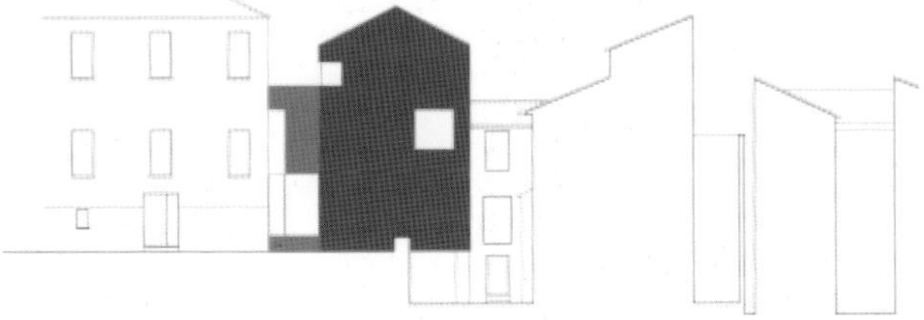

Current construction

Elevations corresponding to different interventions in the main façade across time. The different proposals all opted for adding big windows.

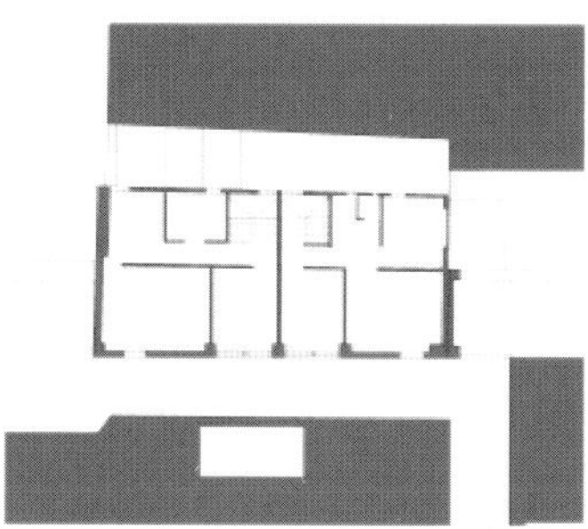

Existing four floor plan

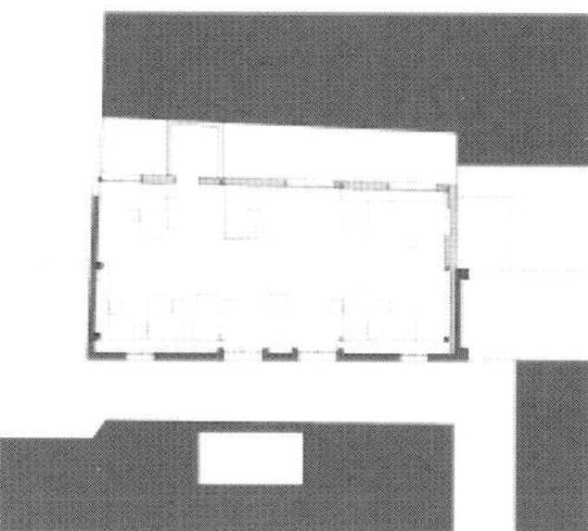

New four floor plan

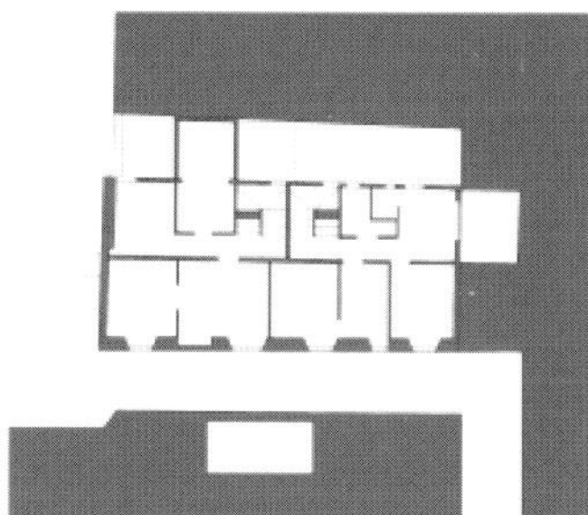

Existing third floor plan

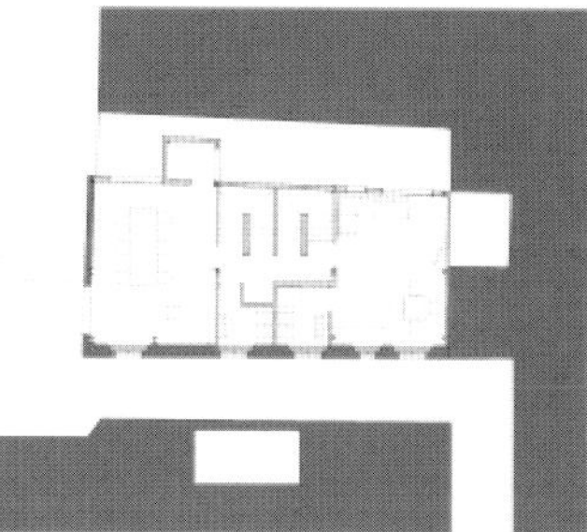

New third floor plan

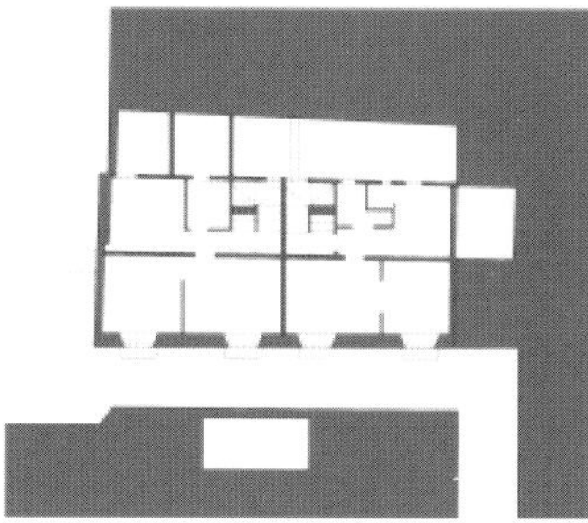

Existing second floor plan

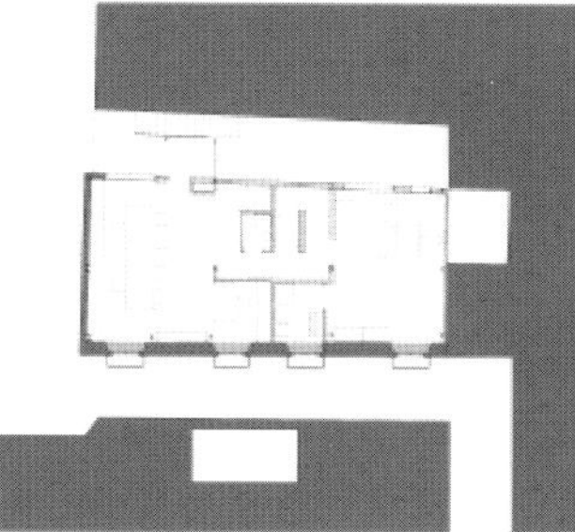

New second floor plan

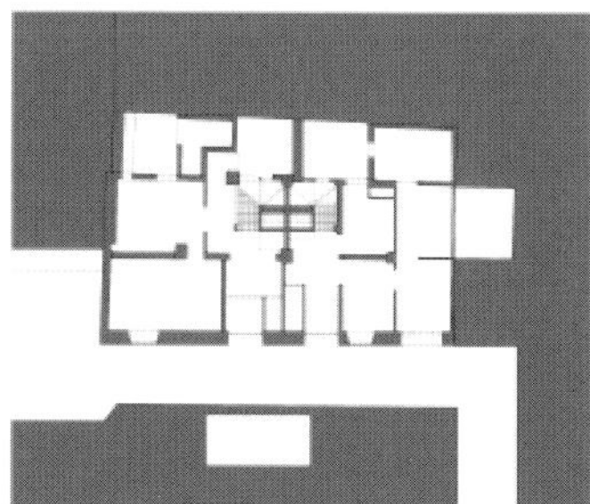

Existing ground floor plan

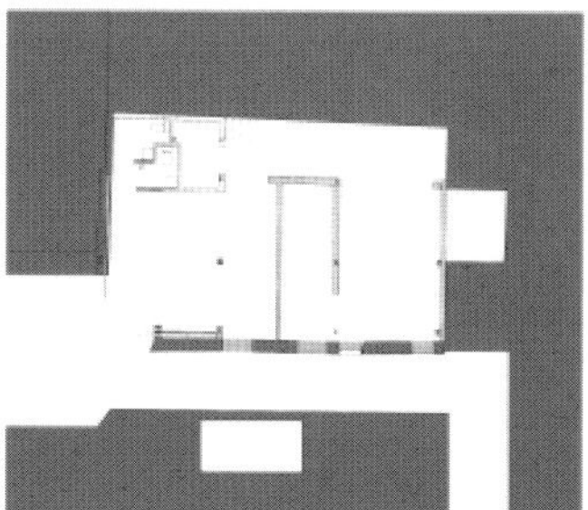

New ground floor plan

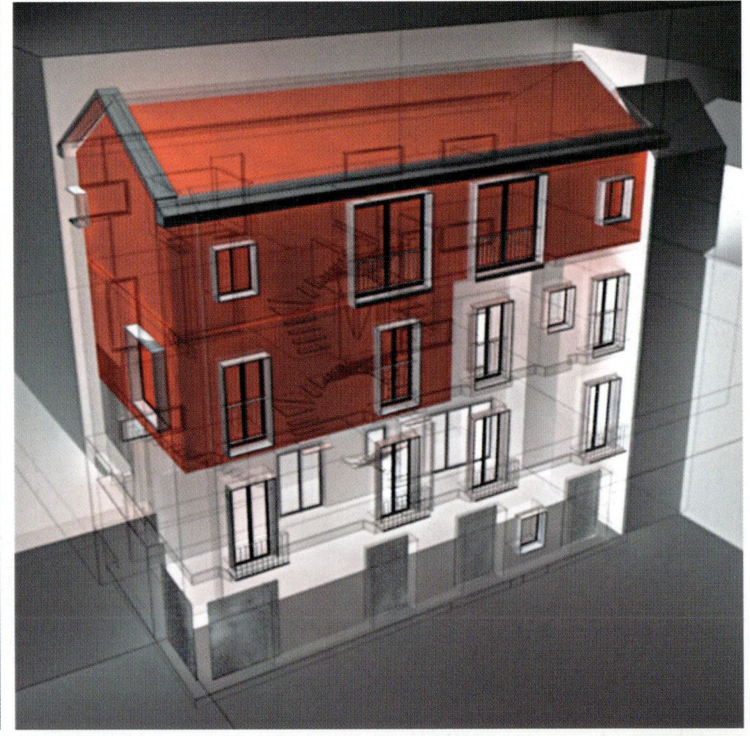
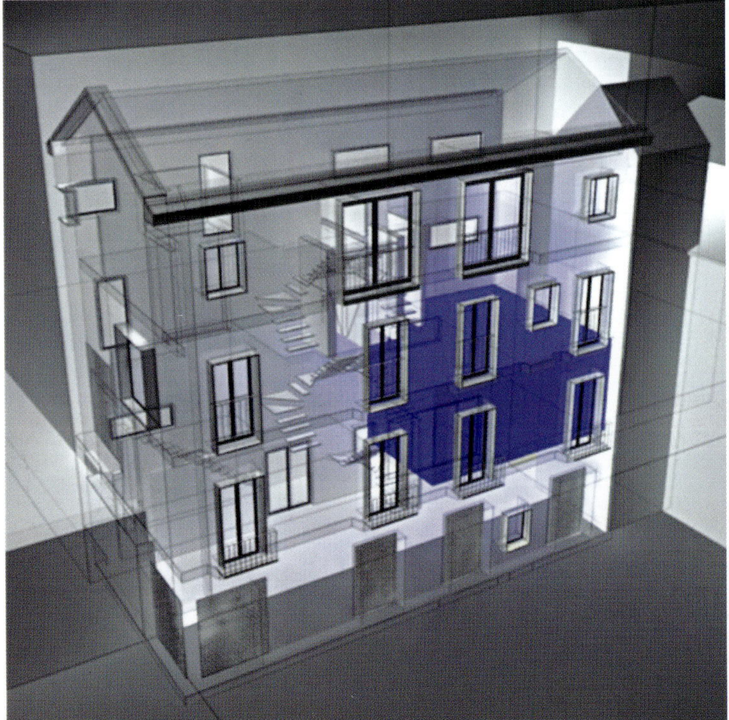
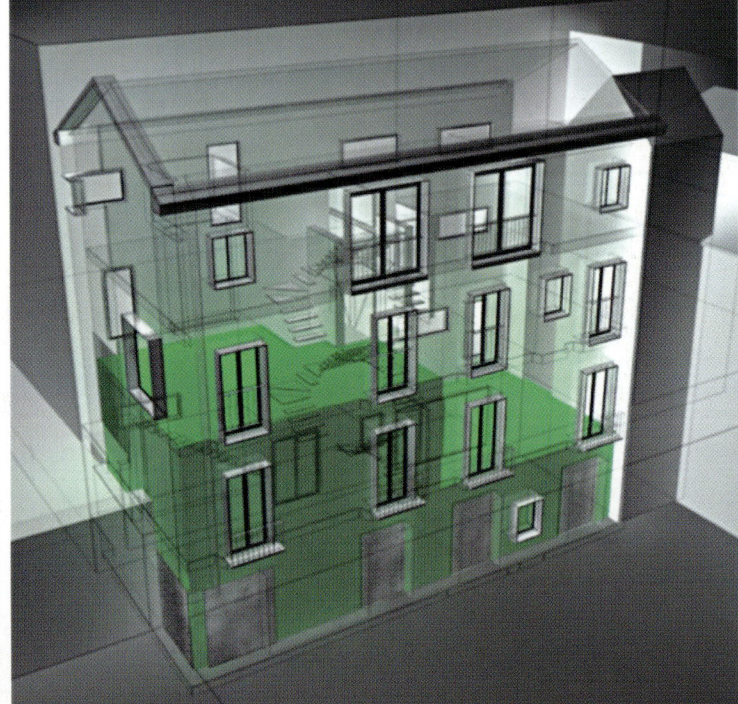

(Previous page). Three-dimensional diagrams showing the layout of the interior spaces. The restaurant is located in the first and ground floor (green). The two apartments are located in the first (blue) and second floors (ochre). The study-office is located in the second and third floors (red).

The interior spaces are connected with the exterior through a window in the side façade. This ruptures the large smooth façade that was unconnected with its surroundings. This big square window provides the office spaces contained within with ventilation and daylighting.

CONJUNTO 2GB

> Mexico City, Mexico | 2003 | Duration of project: 13 months | 11,302 sq ft | © Sebastián Saldívar <

Garduño Arquitectos

This intervention was carried out on a building whose structure was very dilapidated. The properly was approximately 50 years old and contained three apartments.

The project's objectives were to: increase habitable space; restore the structure, add two parking spaces for each dwelling, create a breakfast bar, TV room and extra bedroom, orientate each room to benefit from maximum daylighting while maintaining privacy; regenerate green areas and plant local species, create functional spaces that adapt to the needs of modern living, and radically change the aesthetic of the original construction, giving it a new image in keeping with contemporary design.

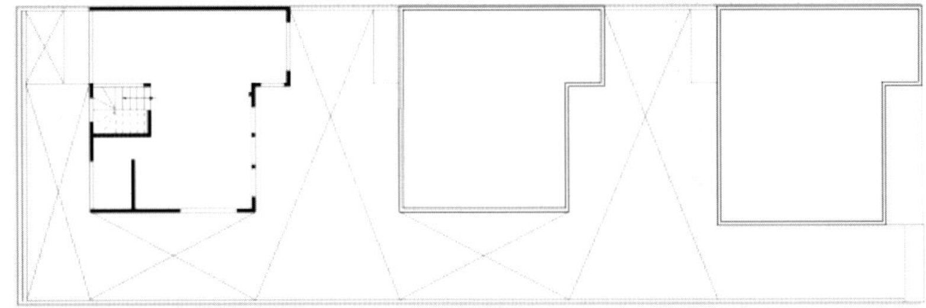

Existing third floor plan

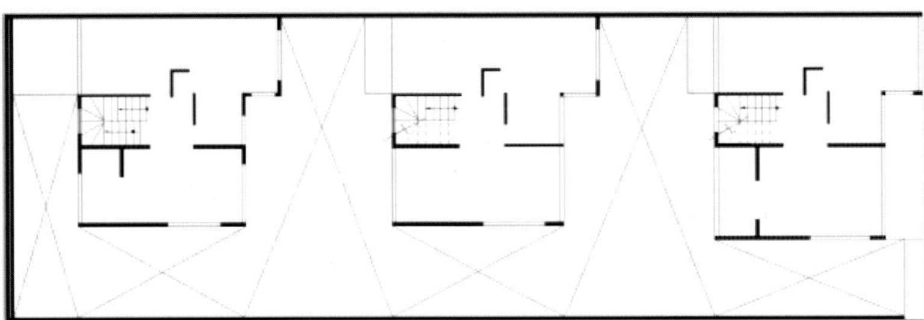

Existing second floor plan

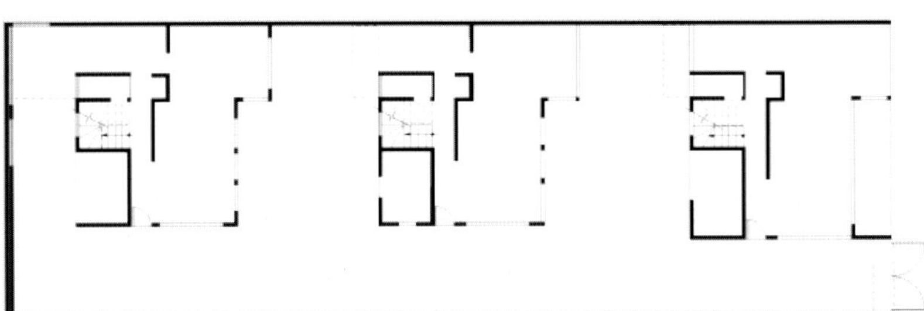

Existing ground floor plan

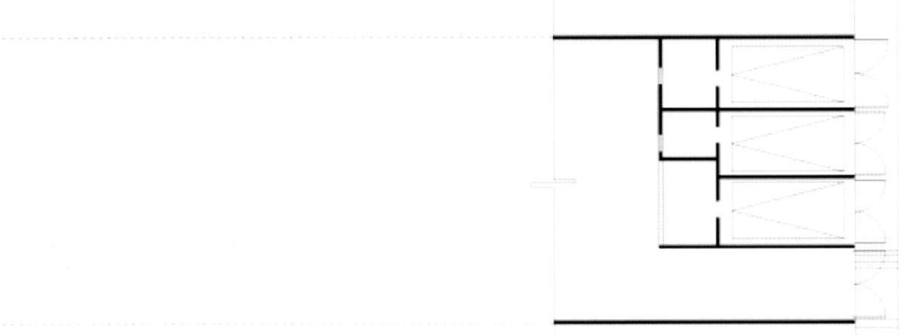

Existing basement floor plan

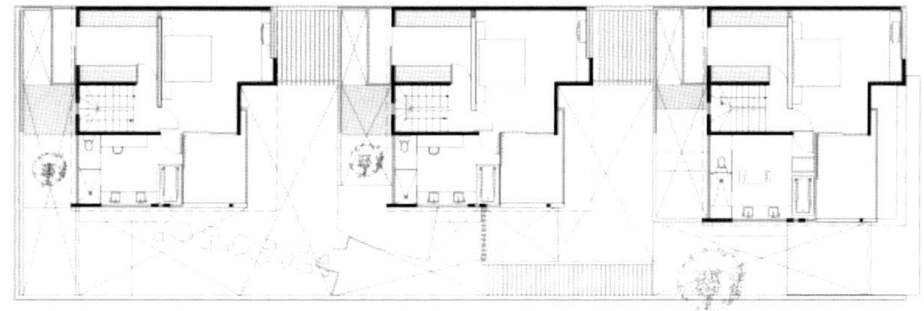

New third floor plan

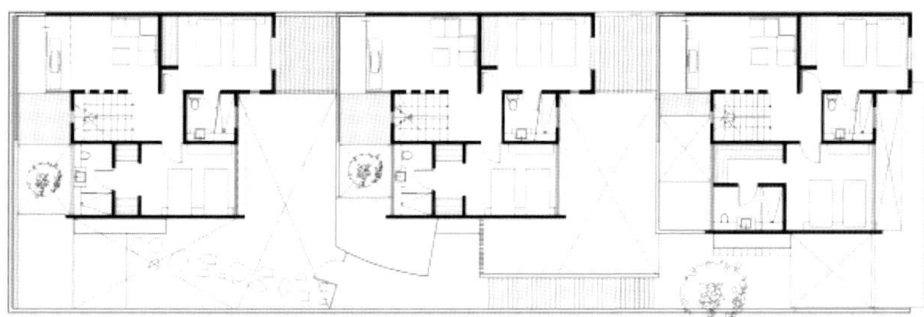

New second floor plan

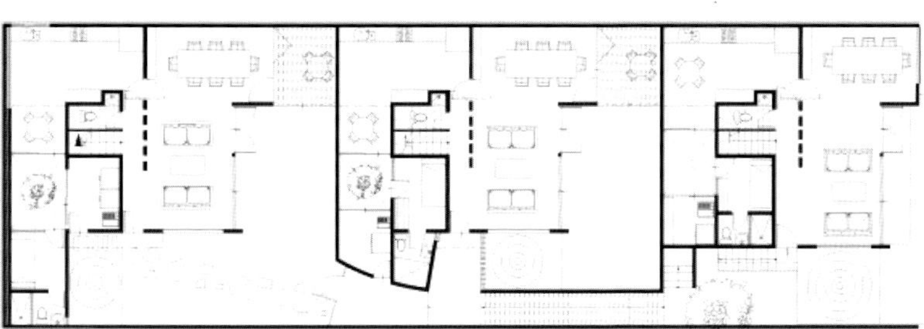

New ground floor plan

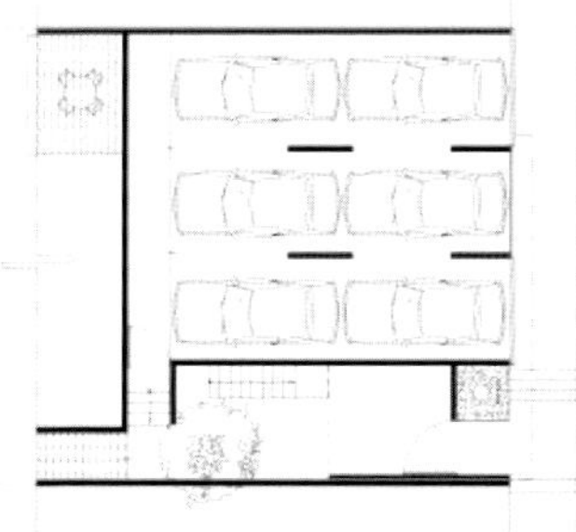

New basement floor plan

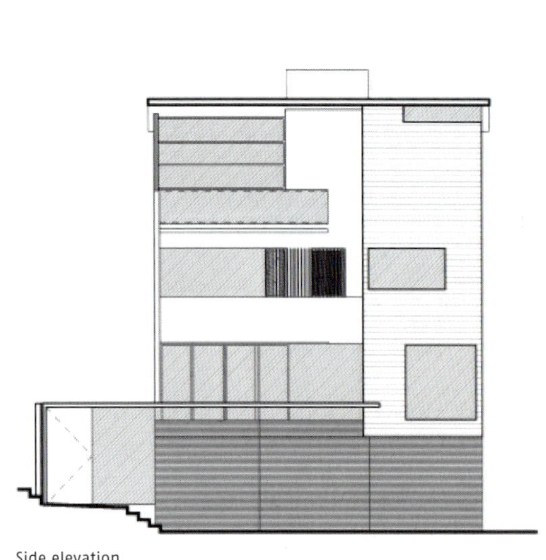

Side elevation

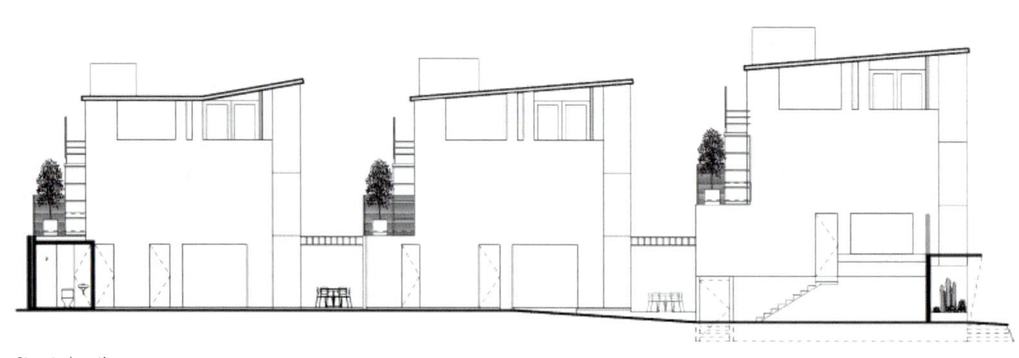

Street elevation

From the outset, the client was involved in the market study and determining consumer trends, the type of housing to constructs and the regulations in the area. The remodeling was staged. The first phase determined the parking space needed, and consolidated and strengthened the foundations to ensure the house was structurally sound.

In the second remodeling phase, loads were replaced out through installing lightweight materials in order to reduce the building's weight and underpin each level of the property. A third level was added in each apartment and the loads were replaced with lightweight materials such as polystyrene roof and wall panels. The project results can be seen in various aspects such as in the building's new image through contrasting traditional and innovative architecture. The main work focused on renovating the structure, aesthetics and functionality.

THE HOUSE OF COLORS

> Sant Just d'Esvern, Spain | 2007 | Duration of project: 17 months | 8,611 sq ft | © Eugeni Pons <

The residential complex was built in 1908 and consisted of a ground floor containing an apartment and commercial premises, two upper floors with six apartments, and a non-accessible roof. This residential program was very limited, and some of the spaces were poorly conditioned, unventilated and lacked daylighting.

The client's objective was an immediate total renovation to convert the building into a solid, habitable and functional residence to be able to rent it out to young people.

The new structure has a much freer spatial program. On the ground floor the commercial premises and apartment were renovated to make them larger, more continual and luminous, through knocking through all the interior walls. In each of the two upper floors contains three types of apartment: a first 377 sq ft apartment, a second one of 484 sq ft, third measuring 6,450 sq ft. The old roof was replaced by a new roof terrace, creating a public space to encourage interaction among neighbors.

Side elevation

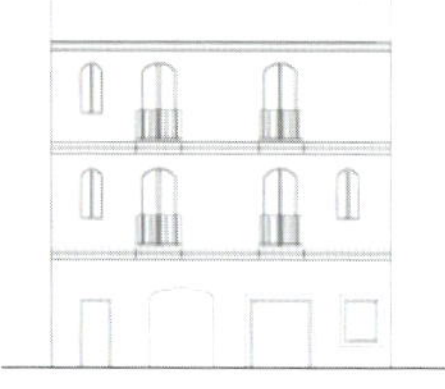

Front elevation

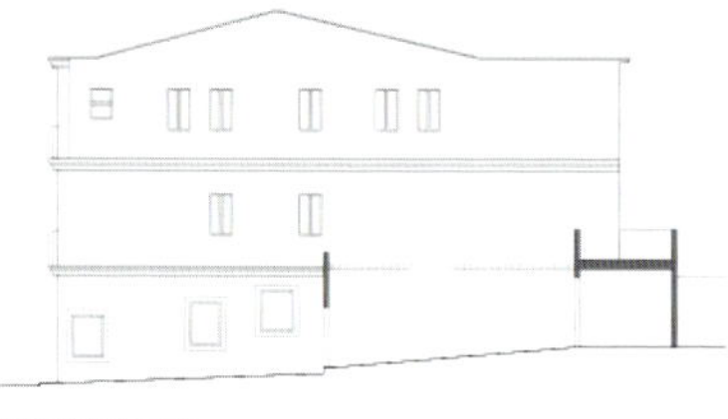

Side elevation

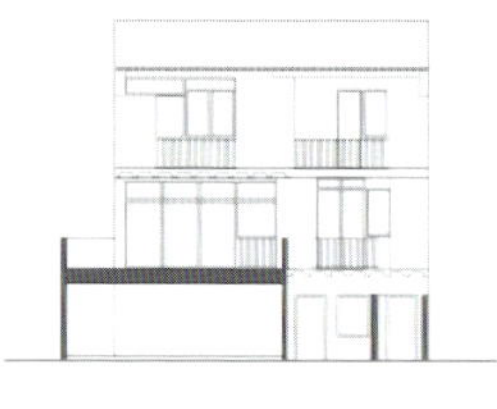

West section perspective

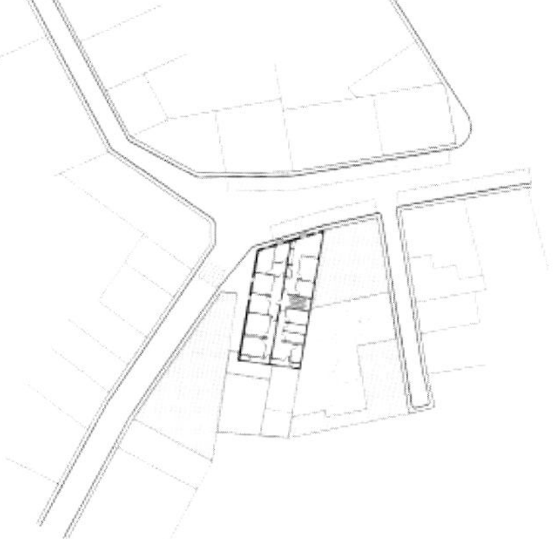

Existing typical floor plan

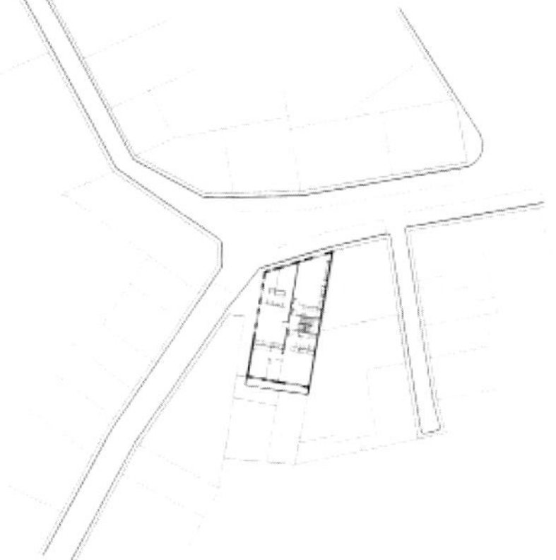

New typical floor plan

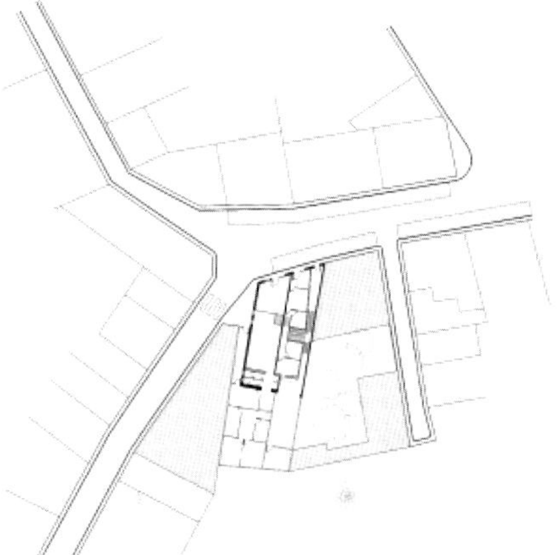

Existing ground floor plan

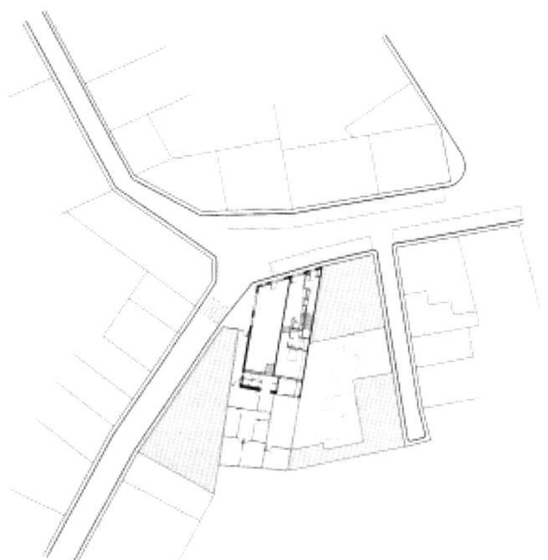

New ground floor plan

The surveyor's study revealed an immediate need to underpin all the property's floors and to demolish the courtyard façade, since there was a danger of immediate collapse. Once the building was stabilized, the property was totally renovated to consolidate its structure and improve its minimal living conditions.

Existing building section

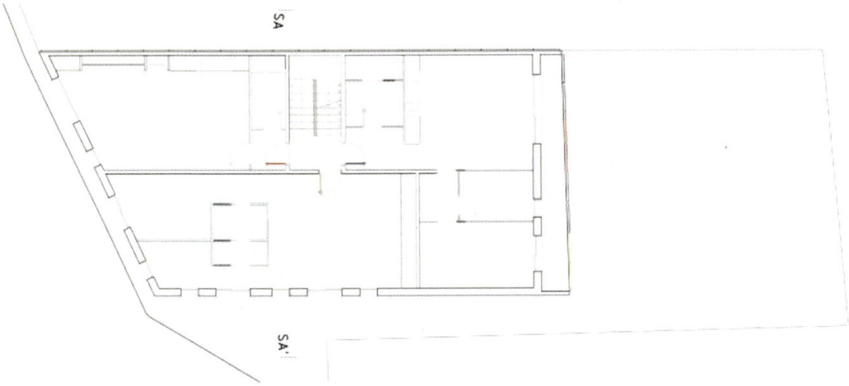

Detail at typical floor plan

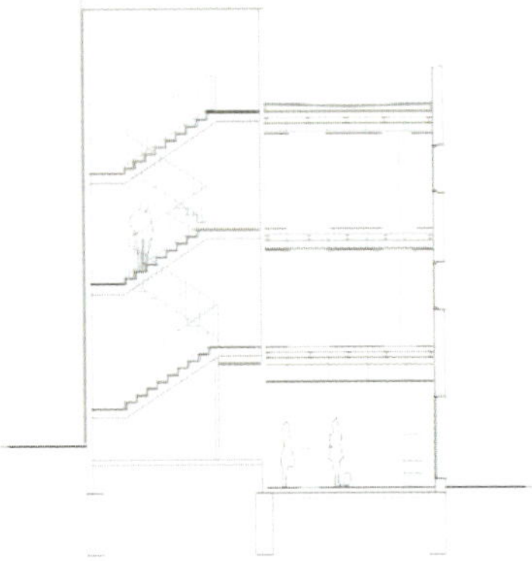

New building section

"Free the exterior walls from their structural function to transform the building into a container that adapts to people's needs"

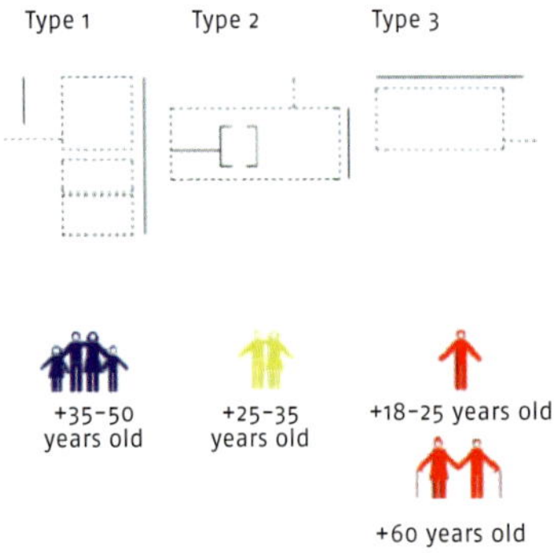

Type 1 Type 2 Type 3

+35–50 years old

+25–35 years old

+18–25 years old

+60 years old

New side elevation

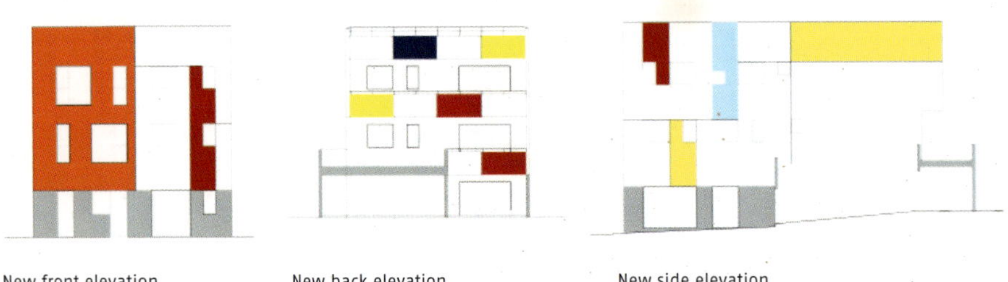

New front elevation

New back elevation

New side elevation

Sustainability and ecological criteria formed a part of the building's renovation project. The new roof terrace was equipped with solar panels and water harvesting tanks. Asymmetrical colored balconies were also added on the new frontage.

The architect's decision to create a new structural organization inside the building was a huge time saver because it enabled a system to be used similar to that used in new builds, not a slow and costly remodeling. This reduced the cost, which was a major concern for the client, and enabled the volume to be stabilized without losing any of existing space, although this entailed creating a project with fewer possibilities. Nevertheless, the result was still more than satisfactory.

> New York, NY, USA | 2007 | Duration of project: 3 years | 40,000 sq ft | © Bjorg Photography <

Chelsea Atelier Architect

This 12-story building is located in the Tribeca district, one of the most modern areas of New York. Most of the apartments are lofts of 1,800 ft² and 3,000 ft².

The architects' intervention consisted in renovating a 19th century apartment block. Prior to the remodeling, the block contained five levels of apartments, and the outer structure was formed by bricks typical of the American architecture of that century. The refurbishment focused on maintaining the structure, add seven more floors and creating a more contemporary interior. Most of the lofts contain a kitchen-dining area, home spa, fireplace, air conditioning, dressing rooms and private terraces. The public spaces for all the neighbors include a lobby, a children's play room and a gym.

The aim of the architect's studio was to create a design that was comfortable but at the same time functional. Due to the speculative nature of the project, this posed a challenge because they had to work with plans that would attract a large number of potential customers. The materials were an important part of achieving this goal. The end result was to determine variations from a minimal selection of these materials.

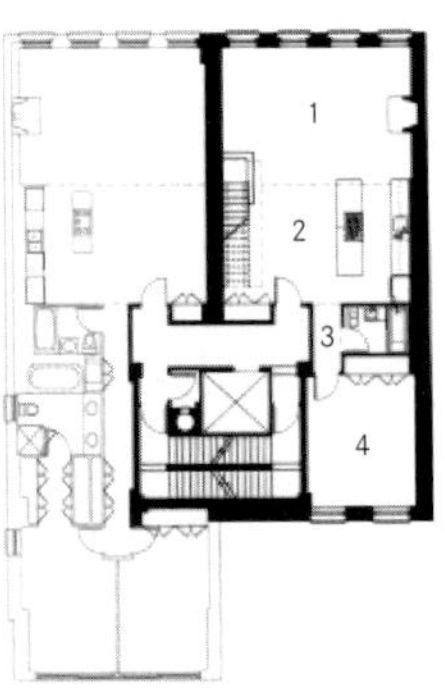

New first floor duplex plan

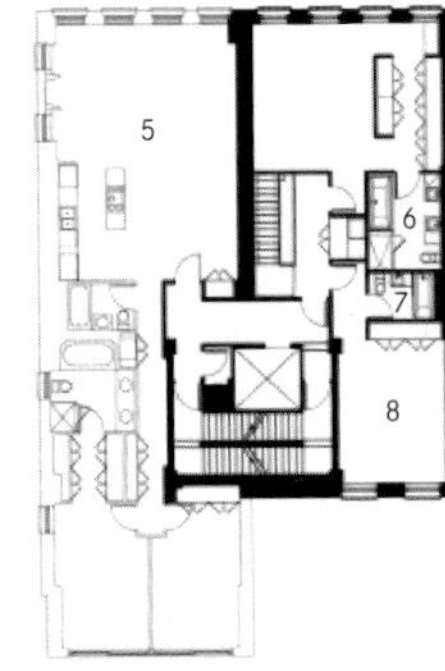

New second floor duplex plan

1. Living room
2. Kitchen
3. Bathroom 1
4. Bedroom 1
5. Master bedroom
6. Master bathroom
7. Bathroom 2
8. Bedroom 2

Existing third–fourth floor plan

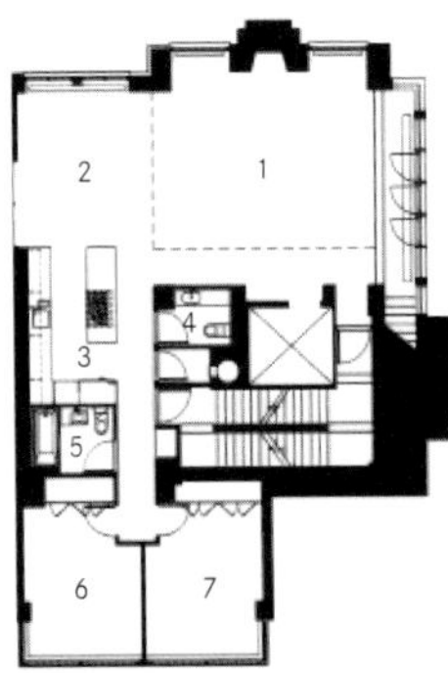

New first floor penthouse plan

New eleventh floor plan

1. Living room
2. Dining room
3. Kitchen
4. Bathroom 1
5. Bathroom 2
6. Bedroom 1
7. Bedroom 2
8. Master bedroom

9. Living room
10. Dining room
11. Master bathroom
12. Bathroom 3
13. Bedroom 3
14. Bedroom 4
15. Bathroom 4
16. Kitchen

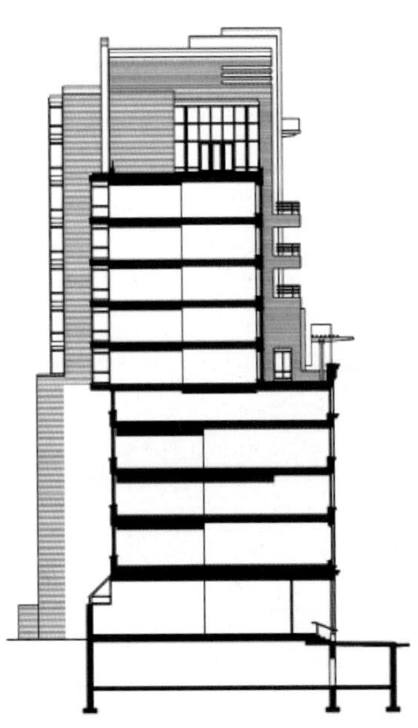

Building section

The kitchens were designed as islands to enhance the idea of a single space, characteristic of lofts. These islands provide a work surface for preparing and cooking food and are finished in bright white. All kitchens include stainless steel appliances and spacious cabinets for storage.

13 DE SEPTIEMBRE HOUSES

JS² Diseño Desarrollo

> Mexico City, Mexico | 2004 | Duration of project: 1 year | 46,284 sq ft | © T. Casademunt, L. Gordoa, J. Navarrete <

This former union warehouse was remodeled into 37 small apartments. This social housing complex is situated in the vast metropolis of Mexico City. The project consisted in keeping the superstructure of the building, restoring it, and demolishing and completely rebuilding the remaining structures. The partition walls were removed, and a new roof and façades were added. The rest of the buildings are organized around an open central courtyard that formed part of the original building. A community space has been created in this courtyard.

Of the total built area of 46,300 sq ft, the houses are distributed in houses of 645 and 1290 sq ft. All apartments are two-levels and are accessed through stairs located in the central courtyard. The design was contemplated to emphasize the natural light entering the courtyard and through the windows in the façades.

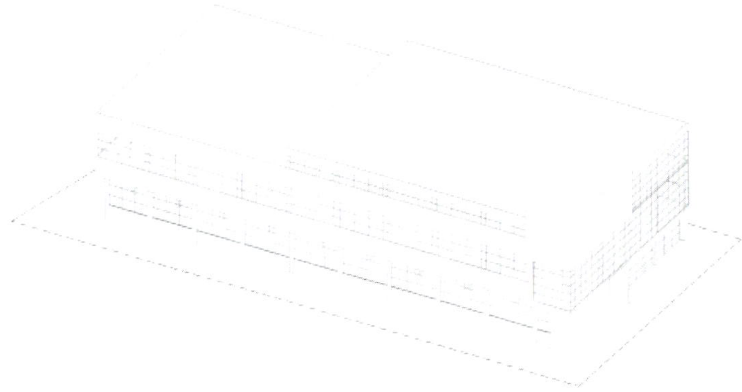

Axonometric – existing conditions

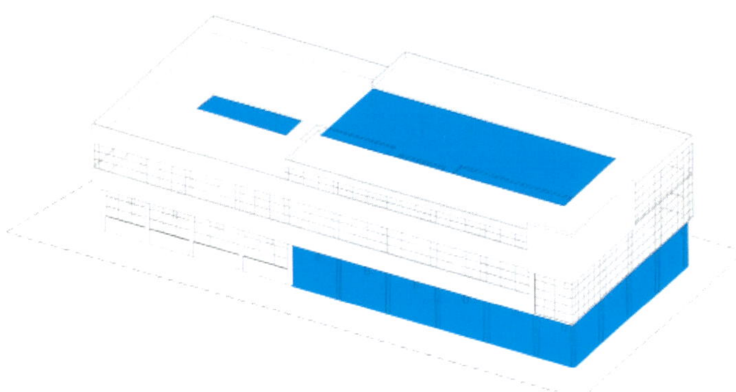

Axonometric – new proposal

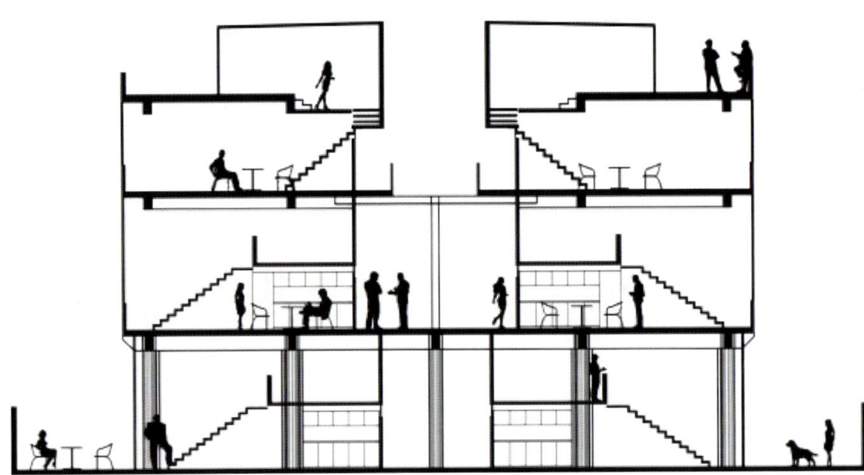

Cross section

When carrying out the remodeling project, the architects decided not to enclose the central courtyard of the old warehouse. This open space provides occupants with windows and natural lighting. The effect of the sunlight on small porches creates an interplay of light and shade on the walls.

Between the first floor and the basement, the architects enclosed the courtyard and installed a cement floor. At either end of the courtyard there are entrances that lead down to the basement, on one side, and, on the other, to the second level of apartments. In the basement, the architects designed parking areas.

The materials used in this intervention were steel for the woodwork and handrail fixtures, aluminum for exterior cladding, and timber for the walkways and the flooring. The two longitudinal walkways create open transit spaces at the ends of the courtyard, and two transversal connections run between them.

> A Coruña, Spain | 2008 | Duration of project: 2 years | 3,122 sq ft | © Franco Pagnozzi Luaces <

This detached house is part of a row of houses built in the Las Flores district. This well-known area of the Galician city was built in the last rationalist phase at the close of the 1960s. The key concept underpinning the renovation was respecting the original spirit of the era in which the house was built, especially the interior layout of spaces.

The client wanted a home that was also a workshop. The main floor is designed as an open-plan space that is large enough to accommodate all the functions in different spaces. The spaces are connected by large sliding panels of tempered glass and steel; these can be placed in different positions to suit the needs of each moment. These divisions are directly inspired by Japanese houses, characterized by versatility and functionality.

The house consists of a ground floor for common use, a first floor containing the private spaces, and a basement where there is a home spa. The design contains a variety of colors such as pearl gray shades that reveal the nuances in the lighting in the living room, dining room and kitchen; camel and gray shades on the bedroom walls; and finally, slate green for the wet areas (see earlier photos).

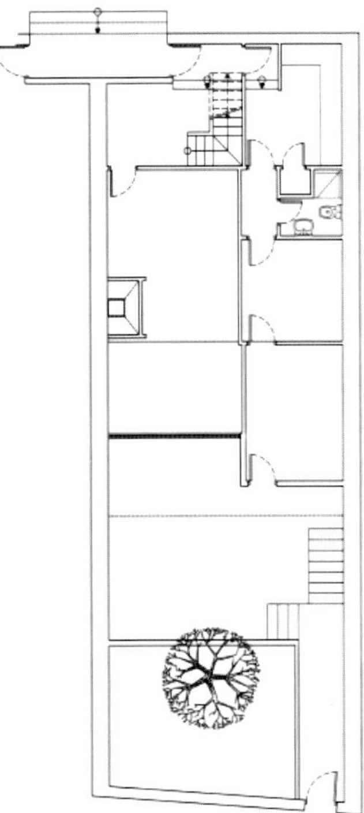

Existing floor plan type A

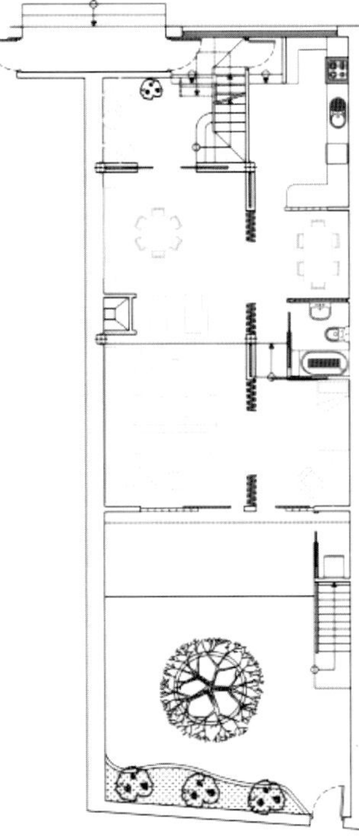

New floor plan type A

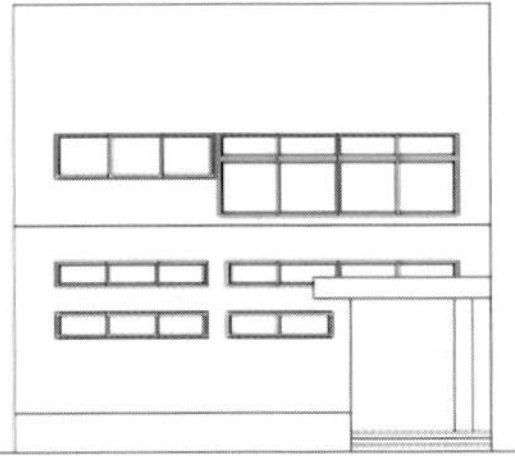

Existing street elevation type A

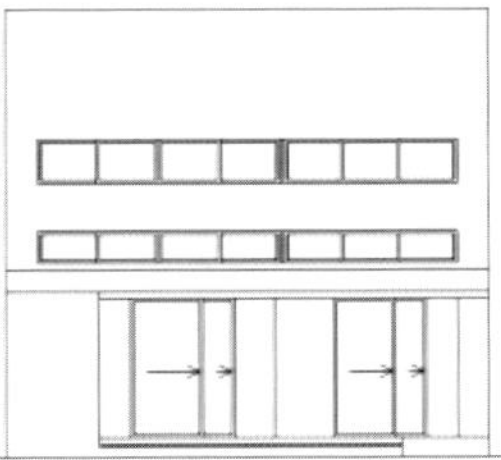

New garden elevation type A

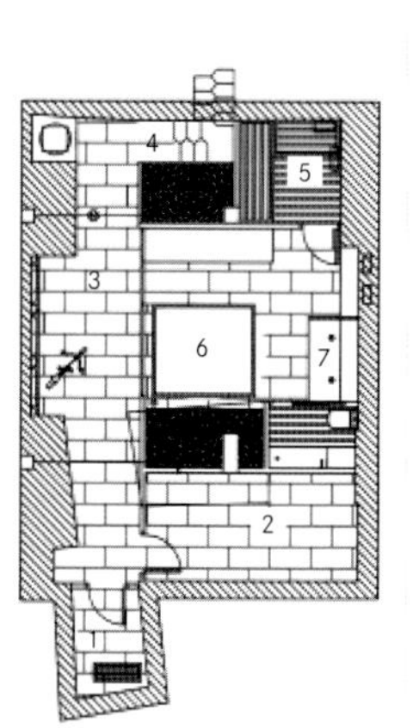

New basement plan type B

1. Ventilation
2. Cellar
3. Gym
4. Japanese garden
5. Sauna
6. Jacuzzi
7. Shower

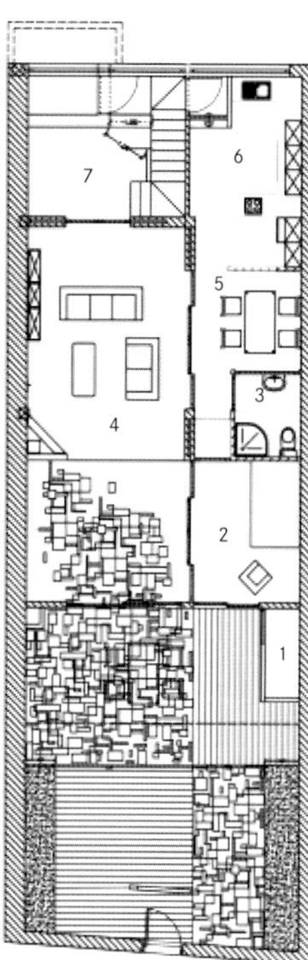

New ground floor plan type B

1. Laundry room
2. Studio
3. Bathroom
4. Living room/dining room
5. Dining room
6. Kitchen
7. Hall

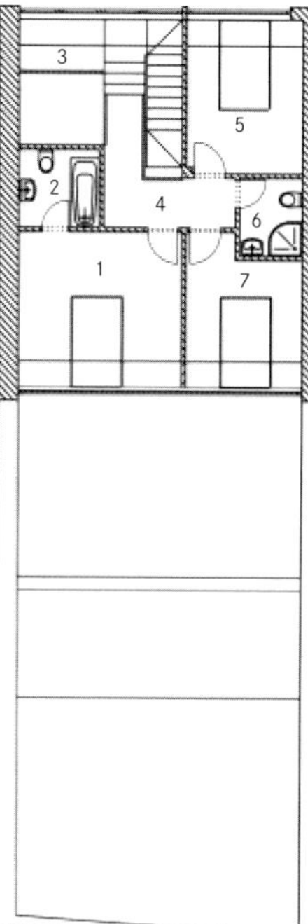

New second floor plan type B

1. Master bedroom
2. Bathroom 1
3. Reading room
4. Hallway
5. Bedroom 1
6. Bathroom 2
7. Bedroom 2

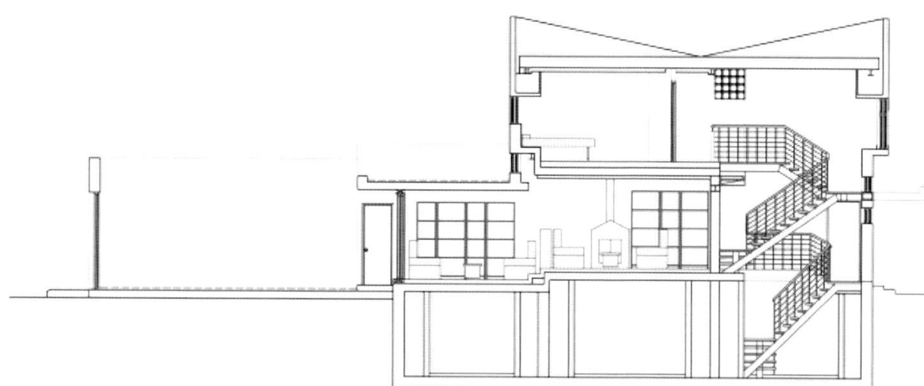

Longitudinal section type B

As well as glazed partitions, other features help define the spaces, such as the effect created by different heights and the vertical accesses. The hall connects the ground floor with the first floor, where the night time double-height spaces are located. In this area, a cannon of light beams through a gallery. The staircase was treated as a sculptural piece, and its lightness allows the light to glide on the glass treads and metal structure.

The basement floor was an addition to the original layout. The main function of this space was to create an area for rest and relaxation. It was for this reason that a home spa was installed there with intimate lighting and a cozy atmosphere, which is just perfect for winding down. This was achieved by installing indirect lighting and tempered glass panels. Miniature Japanese gardens were used in the decoration as a subtle reference to Zen culture.

The floorings used consisted of imitation 1 x 2 ft grayish green slate on the ground floor; the same material in a rusty tone in the patio-garden; wenge combined with oak in horizontal stripes on the first floor; gray imitation slate cut and trimmed ceramic stoneware in the bathroom floors; finally, in the basement the floors are imitation greenish gray slate, and Brazilian walnut in the sauna and Jacuzzi.

Natural lighting is specifically channeled to maximize its effect. With respect to the artificial lighting, light boxes inspired in the cuts of the façade have been installed, therefore creating background of indirect lighting, contrasted by spotlights that create a space that is especially dramatic by night. Halogen lamps are embedded in the pavement, and contained in metal boxes on the walls. Both the mural in the entrance and the furniture and flooring use pop art details that provide a variety of colors.

BOB361 Architects

> Brussels, Belgium | 2007 | Duration of project: 4 years | 30,526 sq ft | © BOB361 Architects <

This building is located along a ring road in the city of Brussels. It was remodeled by a Belgian architect's studio. The building was formerly a goldsmith's workshop with architectural characteristics typical of a factory, such as low ceilings, a red brick façade, and big arc-shaped windows.

The project commissioned was to transform this building into a space for working and living. The architects wanted to maintain the industrial-looking façade and totally modify the interior. The first major action was to partially demolish the extensions that had been added over the years. This increased permeability, since the spaces benefited from better natural lighting and ventilation.

The end result was a property defined by various degrees of privacy: public circulation spaces, semi-private patios, and private lofts and gardens.

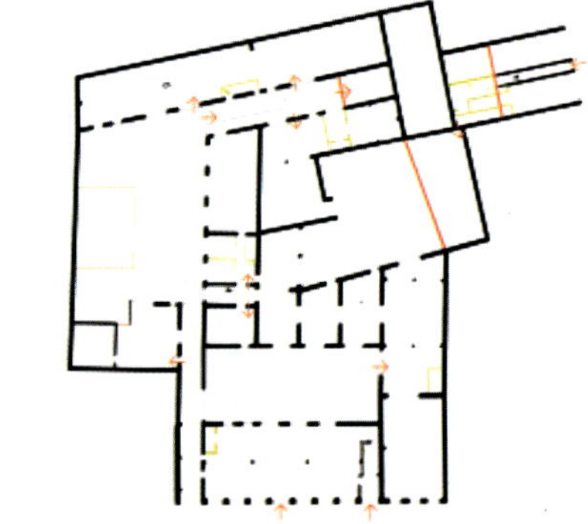

Existing ground floor plan

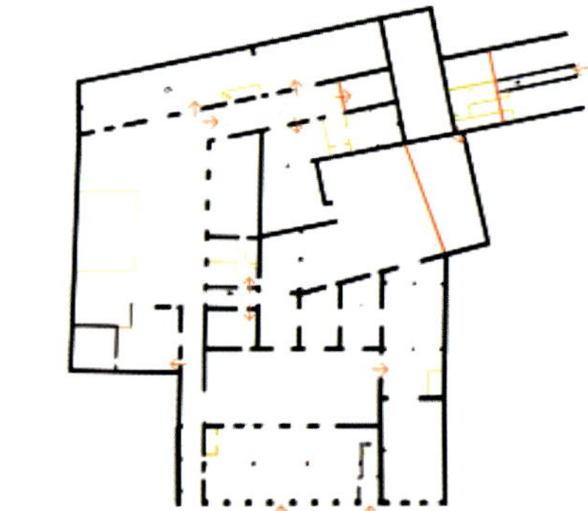

Existing second floor plan

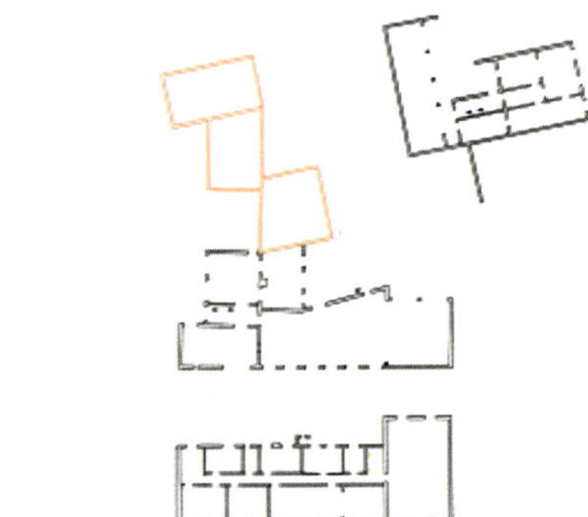

Existing third floor plan

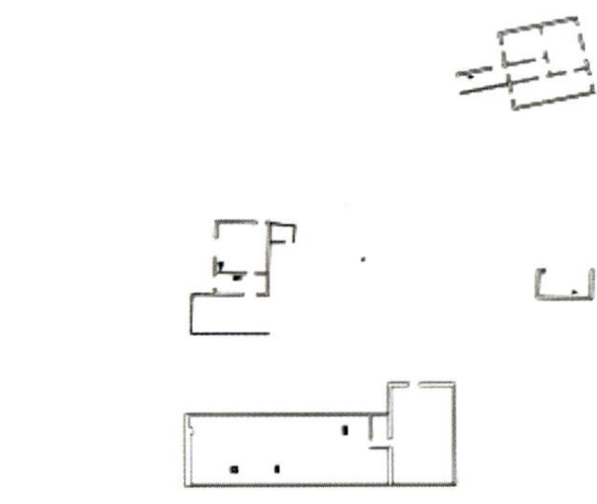

Existing fourth floor plan

Existing building section

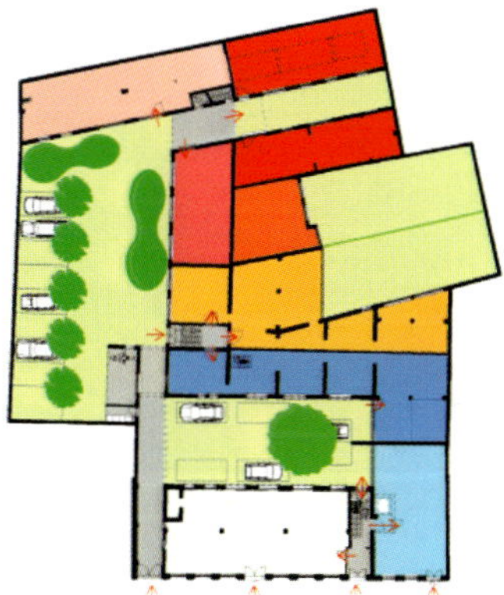

New ground floor plan

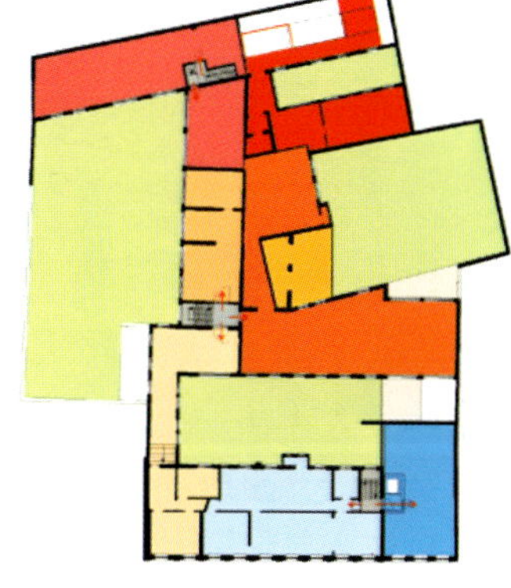

New second floor plan

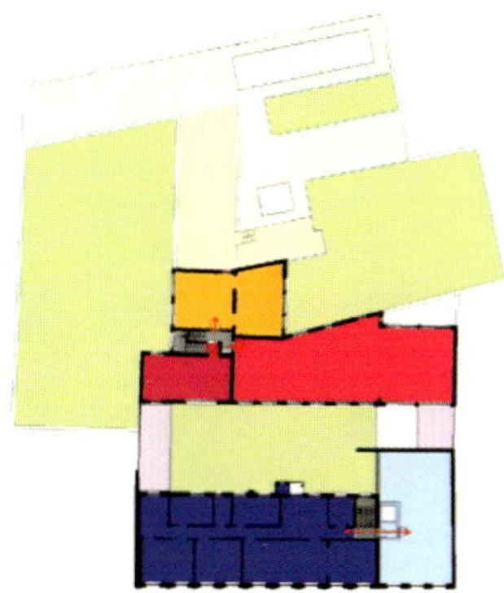

New third floor plan

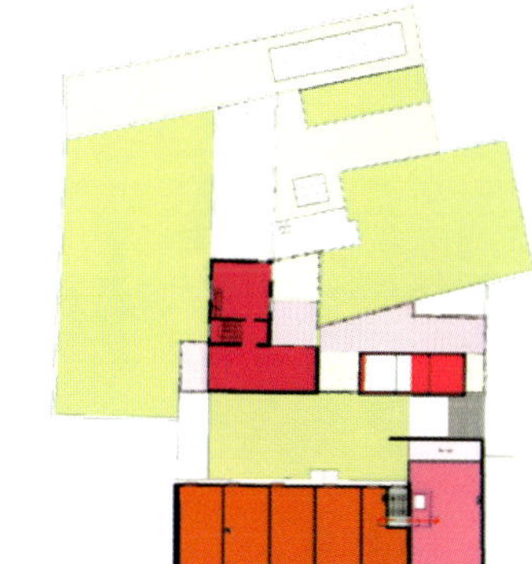

New fourth floor plan

New building section

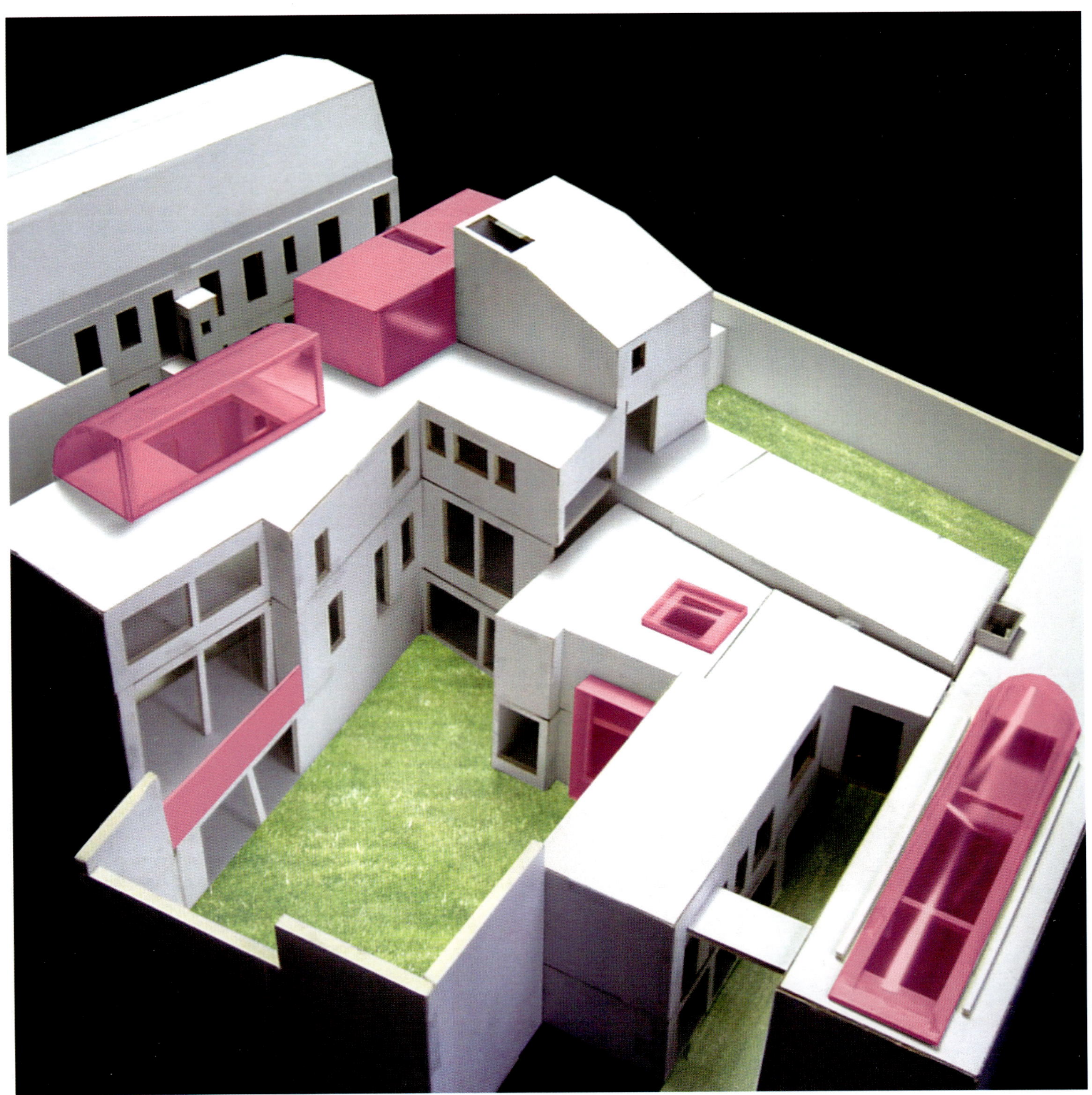

The outer appearance of this former goldsmith's workshop remained practically intact. The architects aimed to maintain the architecture typical of a large factory, in contrast to the interior remodeling. The reorganization included the transforming the transit areas between the different buildings into private gardens.

Each apartment, whether a work space or a dwelling, has a terrace that is accessed through large glazed doors that give access to the exterior and allow daylighting. In this remodeling, the interior spaces gained ventilation and natural light.

Some parts of the building were demolished to free up more space to locate the homes and offices. The architects chose a design where pale colors prevail in the floors and on the walls. This color scheme emphasizes the spaciousness which, coupled with the big windows, creates a bright and warm atmosphere.

DIRECTORY

>Page 250
X-LOFT
Architect: Carola Vannini
Architecture
Rome, Italy
www.carolavannini.com

General contractor: Arcangeli

>Page 258
JM ZARAGOZA APARTMENT
Architect: Magén Arquitectos
Zaragoza, Spain
www.magenarquitectos.com

Principals in charge: Jaime Magén
Pardo, Francisco J. Magén Pardo
Building contractor: M3

>Page 268
GRAPH
Architect: Apollo Architects &
Associates
Chiyoda-ku, Tokyo, Japan
www.kurosakisatoshi.com

>Page 274
**APPARTMENT IN AMERICO VESPUCIO
AVENUE**
Architect: Enrique Browne
Santiago de Chile, Chile
www.ebrowne.cl

Collaborators: Hernán Fontaine, Agustín
Infante, Nicolás Saieh
Constructor: Empresa Constructora
Mena e Infante

>Page 284
ALUMINUM HOUSE
Architect: Hideki Yoshimatsu &
Archipro Architects
Tokyo, Japan
www.archipro.net

Project architects: Hideki Yoshimatsu,
Michio Maeda
General contractor: Shonan Renovation

>Page 294
PRIVATE HOUSE IN FLORENCE
Architect: Paolo Cesaretti
Milan, Italy
www.paolocesaretti.it

Team: Laura Danieli, Valeria Ioele
Interior decoration: Renzo Donnini

>Page 298
HOUSE IN PORTA DA PENA
Architect: RVR Arquitectos
Santiago de Compostela, Spain
www.rvr-arquitectos.es

>Page 304
J-LOFT
Architect: Plystudio
Singapore, Singapore
www.ply-studio.com

General contractor: Paul Renovation
Contractor
Carpentry: Zhongheng Furniture

>Page 314
RESIDENTIAL CONTAINERS
Architect: Petr Hájek/HŠH
Architekti
Prague, Czech Republic
www.hsharchitekti.cz

Contractors: CBT04, Fas Maniny,
Nevšímal, Čermák Truhlářství, Prošek
Zbečno, My Dva Holding

>Page 320
L RESIDENCE
Architect: Min | Day
San Francisco, CA, USA
www.minday.com

Partners in charge:
E.B. Min, Jeffrey L. Day
Collaborators: Christina Kaneva, Natalie
Kittner, Jeff Davis
General contractor: Boyd-Jones
Construction
Structural engineer: Shaffer & Stevens, PC
Lighting design: Min | Day

>Page 330
CHELSEA DUPLEX PENTHOUSE
Architect: Paul Cha Architect
New York, NY, USA
www.paulchaarchitect.com

Project team: Paul Cha, Margaret
Innerhofer, Ali Turan Koluman, Sayoko
Takahashi, Cheng-Ling Lin
Contractor: Halcrow Contracting

>Page 340
LOSA LOFT
Architect: Aidlin Darling Design
San Francisco, CA, USA
www.aidlin-darling-design.com

Principal designers: Joshua Aidlin,
David Darling
Project designer/manager: Ethen Wood
Design team: Michael Pierry
General contractor: McVay Construction
Principal: Monty Montgomery

>Page 346
CHIHUAHUA 78 HOUSES
Architect: JSª Diseño Desarrollo
Mexico City, Mexico
www.jsadd.com

Team: Javier Sánchez, Juan Manuel
Soler, Iris Sosa, Mariana Sánchez,
Edurne Turcott, Julián Bernabé
Structure: Grupo SAI, Fernando Valdivia
Engineering: Grupo SAI, Francisco
Castañeda

>Page 354
SAN MATÍAS HOUSES
Architect: Juan Diego López-
Arquillo/LAA Arquitectos
Granada, Spain
www.laa-arquitectura.com

Quantity surveyor: Juan Gabriel Parilla
Bogas
Collaborators: David Molina Carneros,
Dunia Morilla Radial, Carmen Reina
López
Contractor: Construcciones Salmerón
Monedero, S.L.

CONJUNTO 2GB
Architect: Garduño Arquitectos
Mexico City, Mexico
www.gardunoarquitectos.com

Team: Juan Garduño, Ernesto Flores,
Ricardo Guzmán, Sergio Sánchez
Constructor: María Teresa Rivera, Toyan
Construcciones

THE HOUSE OF COLORS
Architect: Joan Casals Pañella/CSLS
Arquitectes
Hospitalet de Llobregat, Spain
www.csls.es

Contractor: Joan Casals Navarro
Woodwork: Joaquín Carretero (outside),
Nova Porta (inside)

92 WARREN HOUSE
Architect: Chelsea Atelier Architect
New York, NY, USA
www.chelseaatelier.com

Principal in charge: Ayhan Ozan
Project architect: Michael Diani
Collaborator: Manuel Glas
Structural engineer: The Cantor Seinuk
Group, PC
MEP: IV Consulting Engineers
General contractor: Onekey LLC
Stone installer: Krunfol Marble

13 DE SEPTIEMBRE HOUSES
Architect: JSª Diseño Desarrollo
Mexico City, Mexico
www.jsadd.com

Team: Javier Sánchez, Juan Reyes, Iris
Sosa, Mariana Mas, María Elena Reyes
Construction: Waldo Higuera, Hugo
Rivera

**BARRIO LAS FLORES SINGLE FAMILY
HOUSE**
Architect: Miguel Gay Puente
Asturias, Spain
migogay@hotmail.com

P.NT2 HOUSE
Architect: BOB361 Architects
Brussels, Belgium
www.bob361.com

Team: Goedele Desmet, Ivo Vanhamme,
Jean-Michel Culas

Note:
All the drawings and the photographs
taken prior the renovations have been
kindly provided by the architects.